Computational Intelligence-based Time Series Analysis

RIVER PUBLISHERS SERIES IN AUTOMATION, CONTROL AND ROBOTICS

Series Editors:

ISHWAR K. SETHI
Oakland University, USA

TAREK SOBH
University of Bridgeport, USA

FENG QIAO
Shenyang JianZhu University, China

The "River Publishers Series in Automation, Control and Robotics" is a series of comprehensive academic and professional books which focus on the theory and applications of automation, control and robotics. The series focuses on topics ranging from the theory and use of control systems, automation engineering, robotics and intelligent machines.

Books published in the series include research monographs, edited volumes, handbooks and textbooks. The books provide professionals, researchers, educators, and advanced students in the field with an invaluable insight into the latest research and developments.

Topics covered in the series include, but are by no means restricted to the following:

- Robots and Intelligent Machines
- Robotics
- Control Systems
- Control Theory
- Automation Engineering

For a list of other books in this series, visit www.riverpublishers.com

Computational Intelligence-based Time Series Analysis

Editors

Dinesh C. S. Bisht

Jaypee Institute of Information Technology, India

Mangey Ram

Graphic Era Deemed to be University, India
Peter the Great St. Petersburg Polytechnic University, Russia

LONDON AND NEW YORK

Published 2022 by River Publishers
River Publishers
Alsbjergvej 10, 9260 Gistrup, Denmark
www.riverpublishers.com

Distributed exclusively by Routledge
4 Park Square, Milton Park, Abingdon, Oxon OX14 4RN
605 Third Avenue, New York, NY 10158

First published in paperback 2024

Computational Intelligence-based Time Series Analysis / by Dinesh C. S. Bisht, Mangey Ram.

© 2022 River Publishers. All rights reserved. No part of this publication may be reproduced, stored in a retrieval systems, or transmitted in any form or by any means, mechanical, photocopying, recording or otherwise, without prior written permission of the publishers.

Routledge is an imprint of the Taylor & Francis Group, an informa business

Publisher's Note
The publisher has gone to great lengths to ensure the quality of this reprint but points out that some imperfections in the original copies may be apparent.

ISBN 978-10-0079-381-9 (online)

While every effort is made to provide dependable information, the publisher, authors, and editors cannot be held responsible for any errors or omissions.

ISBN: 978-87-7022-417-8 (hbk)
ISBN: 978-87-7004-257-4 (pbk)
ISBN: 978-1-003-33767-6 (ebk)

DOI: 10.1201/9781003337676

Contents

Preface   xi

List of Figures   xiii

List of Tables   xix

List of Contributors   xxi

List of Abbreviations   xxv

1 On Dimensionless Dissimilarity Measures for Time Series   1
K. N. Makris, A. Karagrigoriou, and I. Vonta
- 1.1 Introduction 1
- 1.2 Classical Dissimilarity Measures 3
- 1.3 Classical Entropy-type Dissimilarity Measures 4
- 1.4 Dissimilarity Measures for Time Series Data 5
  - 1.4.1 Standard Dissimilarity Measures 6
  - 1.4.2 Advanced Dissimilarity Measures 12
- 1.5 Conclusions 15

2 The Classification Analysis of Variability of Time Series of Different Origin   19
Teimuraz Matcharashvili, Manana Janiashvili, Rusudan Kutateladze, Tamar Matcharashvili, Zurab Tsveraidze, and Levan Laliashvili
- 2.1 Introduction 20
- 2.2 Used Datasets and Methods of Analysis 21
- 2.3 Results and Discussions 25
- 2.4 Summary 33

3 A Comparative Study of CNN Architectures for Remaining Useful Life Estimation   37
Rahul Joshi, Satvik Bhatt, Amitkumar Patil, and Gunjan Soni
- 3.1 Introduction 38
- 3.2 CNN Architectures and Hyperparameters 40
- 3.3 Numerical Experiments 42
  - 3.3.1 Dataset 42
  - 3.3.2 Pre-processing 43
    - 3.3.2.1 Labelling 44
    - 3.3.2.2 Scaling of data 44
    - 3.3.2.3 Splitting of data 44
    - 3.3.2.4 Time series to recurrence plots 45
- 3.4 Results 46
- 3.5 Conclusion 47

4 The Analysis of Dynamical Changes and Local Seismic Activity of the Enguri Arch Dam   53
Aleksandre Sborshchikovi, Tamaz Chelidze, Ekaterine Mepharidze, Dimitri Tepnadze, Natalia Zhukova, Teimuraz Matcharashvili, and Levan Laliashvili
- 4.1 Introduction 54
- 4.2 Main Text 56
  - 4.2.1 Methods and Results 56
- 4.3 Conclusion 61

5 Analysis and Prediction of Daily Closing Price of Commodity Index Using Auto Regressive Integrated Moving Averages   65
Bijesh Dhyani, Manish Kumar, Poonam Verma, and Anurag Barthwal
- 5.1 Introduction 66
- 5.2 Literature Review 67
- 5.3 Objectives and Study 69
- 5.4 Data and Methodology 69
- 5.5 Data Decomposition 69
  - 5.5.1 Seasonality 69
  - 5.5.2 Trend 69
  - 5.5.3 Cyclicity 70
- 5.6 Augmented Dicky Fuller (ADF) Test 70
  - 5.6.1 Auto-correlation function (ACF) 71

		5.6.2 Partial Autocorrelation Function (PACF)	73
		5.6.3 ARIMA Model	73
	5.7	Results and Analysis	74
	5.8	Conclusion and Future Work	77

6 Neural Networks Analysis of Suspended Sediment Transport Time Series Modeling in a River System   81

M. Harini Reddy, N. Manikumari, M. Mohan Raju, Dinesh C. S. Bisht, A. Naresh, Harish Gupta, and M. Gopal Naik

6.1	Introduction		82
6.2	Artificial Neural Networks		83
6.3	Hydrological Study Area		85
6.4	Methodology		87
	6.4.1	Mathematics of SRC	89
6.5	Results		89
6.6	Hysteresis of Sediment Transport Process		93
6.7	Conclusions		94
6.8	Acknowledgements		94

7 Ranking Forecasting Algorithms Using MCDM Methods: A Python Based Application

Swasti Arya, Mihika Chitranshi, and Yograj Singh   99

7.1	Introduction		100
7.2	Review of Literature		100
	7.2.1	Analytic Hierarchy Process (AHP)	100
	7.2.2	Technique for Order of Preference by Similarity to Ideal Solution (TOPSIS)	101
	7.2.3	VlseKriterijumska Optimizacija I Kompromisno Resenje (VIKOR)	102
	7.2.4	Time Series Analysis	102
7.3	Error Measurements and Forecasting		103
	7.3.1	Holt-Winter	105
	7.3.2	Autoregressive Integrated Moving Average (ARIMA)	105
	7.3.3	SARIMA	106
	7.3.4	ARIMA integrating Single Judgement Adjustment	106
	7.3.5	ARIMA integrating Collaborative Judgement Adjustment	106
7.4	Multi Criteria Decision Making		107
	7.4.1	Multiple Attribute Decision Making (MADM)	107

		7.4.2	Multiple Objective Decision Making (MODM)	108
	7.5	MCDM Methods		108
		7.5.1	The AHP Method	108
		7.5.2	The TOPSIS Method	110
		7.5.3	The VIKOR Method	112
	7.6	Framework of the Problem		114
	7.7	Implementation Using Python Programming Language		115
		7.7.1	Determining the criteria weights using AHP	115
		7.7.2	Ranking Alternatives using TOPSIS method	116
		7.7.3	Ranking Alternatives using VIKOR method	118
	7.8	Result Analysis		119
	7.9	Conclusion		120
	7.10	Appendix		121

8 Rainfall Prediction Using Artificial Neural Network   **127**
Sunil K. Sahu, N. Kumar Swamy, and Dinesh Bisht

	8.1	Introduction		127
	8.2	Materials and Method		129
		8.2.1	Input and Output Data Selection	129
		8.2.2	Input Data Training	129
		8.2.3	Validation and Testing	129
		8.2.4	Artificial Neural Network Architecture	130
	8.3	Result		131
	8.4	Discussion		135
	8.5	Comparision of ANN Model with Regresion		136
	8.6	Conclusion		140

9 Statistical Downscaling and Time Series Analysis for Future Scenarios of Temperature in Haridwar District, Uttarakhand   **143**
Gaurav Singh, Nitin Mishra, and L. N. Thakural

	9.1	Introduction		144
	9.2	Study Area		145
	9.3	Data Used and Methodology		146
		9.3.1	Data Used	146
		9.3.2	Methodology	147
	9.4	Results and Discussion		147
		9.4.1	Regression Method	148
		9.4.2	Predictor Variables Selection	148
		9.4.3	Calibration and Validation Results	148

		9.4.4 Future Emission Scenarios	152
	9.5	Conclusion	156

Index   **159**

About the Editors   **161**

Preface

The sequential analysis of data and information gathered from past to present is called as time series analysis. Time series data are of high dimension, large size and updated continuously. Times series depends on various factors like trend, seasonality, cycle and irregular dataset. A time series is basically a series of data points well-organised in time. Time series forecasting is a significant area of machine learning. There are various prediction problems that are time-dependent and these problems can be handled through time series analysis. Computational intelligence (CI) is a developing computing approach for the forthcoming several years. CI gives the litheness to model the problem according to given requirements. It helps to find swift solutions to the problems arising in numerous disciplines. These methods mimic from human behaviour. The main objective of CI is to develop intelligent machines to provide solutions to real world problems, which are not modelled or too difficult to model mathematically. This book aims to cover the recent advances in time series and applications of CI for the time series analysis. The projected audience for this book will be scientists, researchers and postgraduate students.

Dinesh C. S. Bisht

Mangey Ram

List of Figures

Figure 1.1	Two time series, n=12 (series 1 in red and series 2 in green).	8
Figure 1.2	Two time series, n=12 (series 1 in red and series 2 in green).	9
Figure 2.1	Influence of noise intensity on calculated KLD (grey columns) and MD (black columns) values of time series of X components of Lorenz system (a) and Henon attractor (b).	25
Figure 2.2	Calculated KLD (grey columns) and MD (black columns) values of (a) comparison of regular and irregular parts of BGP updates time series ATT(1), NTT(2), IIJ(3) and Tinet(4); (b) comparison of regular parts of BGP updates time series recorded at ATT AS with BGP updates time series recorded at Tinet(1), NTT(2) and IIJ(3) AS-es.	26
Figure 2.3	KLD (grey columns) and MD (black columns) values of time series of, (a) IET (normed inter earthquakes times) compared with RNT (normed randomised inter earthquakes times), (b) IED (normed inter event distances) compared with RND (normed randomised inter event distances).	27
Figure 2.4	Calculated KLD (grey columns) and MD (black columns) values of meteorological time series. (1) Original AMaxT compared with randomised RAMaxT. (2) Original AMinT compared with randomised RAMinT. (3) AMaxT compared with AMinT.	29

xiv List of Figures

Figure 2.5 Calculated KLD (grey columns) and MD (black columns) values of comparison between systolic and diastolic blood pressure time series: optimal (1), normal (2), high-normal (3) and hypertension (4) categories. 30

Figure 2.6 Calculated KLD (grey columns) and MD (black columns) values of comparison of optimal blood pressure category with normal (1), high-normal (2) and hypertension (3) categories. Systolic (a), diastolic (b) blood pressure and heart rate (c) time series. 31

Figure 2.7 Calculated KLD (grey columns) and MD (black columns) values for south Caucasian Countries when they compared by datasets of sub-components of IEF for the period from 2014 to 2020 (1–7 in abscissa axis); (a) Armenia versus Azerbaijan, (b) Georgia versus Armenia, (c) Georgia versus Azerbaijan. 32

Figure 2.8 Calculated KLD (grey) and MD (black) values for south Caucasian countries when they compared by datasets of sub-components of DB index in period from 2014 to 2020 (1–7 in abscissa axis); Armenia versus Azerbaijan – triangles, Georgia versus Armenia – circles, Georgia versus Azerbaijan – squares. 33

Figure 3.2a LeNet-5. 41
Figure 3.2b AlexNet. 41
Figure 3.2c VGGNet16. 42
Figure 3.3a Steps in pre-processing. 44
Figure 3.3b,c Regression data to recurrence plots. 45
Figure 4.1 Satellite image of the Enguri dam and reservoir area with locations of the main Ingirishi fault and the branch fault, crossing the dam foundation and Dam foundation displacement datasets around EDITA polygon in period from 1974 to 2020. Dam foundation displacement datasets around EDITA polygon in period from 1974 to 2020. Arrows 1–5 correspond to the start of five periods of fault zone extension. 55

Figure 4.2	MF-DFA analysis of displacements of Enguri Arc Dam foundation.	58
Figure 4.3	MF-DFA analysis of seismic catalogue around Enguri Arc Dam. compare of original interevent sequences with random data.	59
Figure 4.4	Lempel and Ziv algorithmic complexity (LZC) analysis of compare original (grey circles) and randomise (black triangle) seismic data around Enguri Arch Dam.	61
Figure 5.1	Time-series of closing price of the commodity index for the duration from the year 2015 to the year 2021.	70
Figure 5.2	The seasonality of the time-series of closing price of the commodity index for the duration from the year 2015 to the year 2021.	70
Figure 5.3	Auto-correlation function (ACF) plot against the lag values of the time-series of closing price of the commodity index for the duration from the year 2015 to the year 2021.	72
Figure 5.4	Partial auto-correlation function (PACF) plot against the lag values of the time-series of closing price of the commodity index for the duration from the year 2015 to the year 2021.	73
Figure 5.5	Time-series of closing price of the commodity index for the duration from 01-12-2015–30-11-2020.	75
Figure 5.6	The time-series of closing price of the commodity index for the duration between the year 2015 and the year 2021 (Blue line represents previous time-series values used for building ARIMA model and red line represents the forecast generated by the model).	75
Figure 5.7	The difference between actual and the predicted closing prices of the commodity index for the test duration between 19-12-2020–12-2020 (Green line represents the actual closing price and maroon line represents the forecast generated by the model).	76
Figure 6.1	ANN architecture with input, output and weight vectors.	84
Figure 6.2	Peddavagu river with Bhatpalli hydrological gauging station in Godavari river system.	86

Figure 6.3	Real time series and ANN modelled suspended sediment plots of Peddavagu river at Bhatpalli hydrological gauging station.	91
Figure 6.4	Real time series, ANN model time series and SRC time series for Peddavagu river at Bhatpalli.	92
Figure 6.5	Observed and estimated hysteresis of sediment load in Peddavagu river basin.	94
Figure 7.1	Criteria weights as calculated using AHP.	117
Figure 7.2	Line graph of relative closeness coefficients in TOPSIS.	118
Figure 7.3	Line graph for comparison of S_i, R_i and Q_i in VIKOR.	119
Figure 7.4	Python code for weight determination by AHP.	121
Figure 7.5	Python code for TOPSIS.	122
Figure 7.6	Python code for VIKOR.	123
Figure 8.1	Architecture of artificial neural network.	130
Figure 8.2	Back propagation algorithm best validation performances case-01.	132
Figure 8.3	Comparison between actual and predicted value for back propagation algorithm.	133
Figure 8.4	Regression plot in back propagation algorithm.	134
Figure 8.5	Layer recurrent network best validation performances case-04.	135
Figure 8.6	Comparison between actual and predicted value for layer recurrent network.	135
Figure 8.7	Cascaded back propagation best validation performances of case-03.	136
Figure 8.8	Cascaded back propagation Regression.	137
Figure 8.9	Comparison between actual and predicted value for cascaded back propagation	138
Figure 8.10	Average prediction error corresponding to back propagation algorithm, cascaded propagation algorithm and layer recurrent network.	138
Figure 8.11	Prediction error for the back propagation algorithm ANN model.	139
Figure 8.12	Prediction errors from back propagation algorithm and multiple linear regression.	139
Figure 9.1	Map of study area (source: www.mapsofindia.com).	146
Figure 9.2(a)	Monthly average max. temperature for calibration period (1961–1995).	149

Figure 9.2(b)	Monthly average max. temperature for validation period (1996–2005).................	150
Figure 9.3(a)	Monthly average min. temperature for calibration period (1961–1995).................	150
Figure 9.3(b)	Monthly average min. temperature for validation period (1996–2005).................	150
Figure 9.4(a)	Monthly average maximum temperature for climate scenarios 2006–2040.................	152
Figure 9.4(b)	Monthly average maximum temperature for climate scenarios 2041–2070.................	152
Figure 9.4(c)	Monthly average maximum temperature for climate scenarios 2071–2099.................	153
Figure 9.5(a)	Monthly average minimum temperature for climate scenarios 2006–2040.................	153
Figure 9.5(b)	Monthly average minimum temperature for climate scenarios 2041–2070.................	153
Figure 9.5(c)	Monthly average minimum temperature for climate scenarios 2071–2099.................	154

List of Tables

Table 3.1	Selected hyperparameters for networks	42
Table 3.2	Twenty-one sensor recordings with three operational settings	43
Table 3.3	Labelling of dataset	44
Table 3.4	Window-wise segregation	45
Table 3.5	Performance metrics for proposed architectures	46
Table 5.1	The actual commodity price, forecasted price, absolute error and error percent for the time duration from 19-12-2020–27-12-2020	76
Table 6.1	Runoff-sediment models developed for the study	88
Table 6.2	Performance comparison of ANN modeling and SRC method for Peddavagu river at Bhatpalli hydrological gauging station.	90
Table 7.1	Descriptions of error measurements	104
Table 7.2	Ratio scale for AHP	109
Table 7.3	Random index or consistency index for AHP	110
Table 7.4	Best and worst values for criteria for TOPSIS	111
Table 7.5	Best and worst values for criteria for VIKOR	113
Table 7.6	Best and worst values of S_i and R_i	113
Table 7.7	Description of criteria	115
Table 7.8	Description of the alternatives	115
Table 7.9	Decision matrix	115
Table 7.10	Pairwise comparison matrix	116
Table 7.11	Criteria weights calculated using AHP	116
Table 7.12	Weighted normalised decision matrix	117
Table 7.13	Positive ideal solution and negative ideal solution	117
Table 7.14	Ranked alternatives based on TOPSIS	118
Table 7.15	Values of S_i, R_i and Q_i for VIKOR	119
Table 7.16	Ranks based on S_i, R_i and Q_i for VIKOR	119
Table 8.1	Comparison of MSE for different cases using back propagation algorithm	132

Table 8.2	Comparison of MSE for different cases for layer recurrent network	133
Table 8.3	Comparison of MSE for different cascaded back propagation	134
Table 9.1	Selected NCEP predictors and their relationship with maximum temperature	149
Table 9.2	Selected NCEP predictors and their relationship with minimum temperature	149
Table 9.3	Calibration and validation of monthly average maximum temperature (°C) with NCEP reanalysis data months calibration	151
Table 9.4	Calibration and validation of monthly average minimum temperature (°C) with NCEP reanalysis data months calibration	151
Table 9.5	Detailed average maximum temperature (°C) statistics for different time steps (scenarios)	154
Table 9.6	Detailed average minimum temperature (^oC) statistics for different time steps (scenarios)	155
Table 9.7	Average yearly maximum temperature for present and downscaled maximum temperature corresponding to RCPs (2.6, 4.5 and 8.5) scenario	155
Table 9.8	Average yearly minimum temperature for present and downscaled minimum temperature corresponding to RCPs (2.6, 4.5 and 8.5) scenario	156

List of Contributors

Arya, Swasti, *Department of Mathematics, Lady Shri Ram College for Women, University of Delhi, India; E-mail: swasarya@gmail.com*

Barthwal, Anurag, *Management Studies, Graphic Era Deemed To Be University, Management Studies, Graphic Era University, Graphic Era Hill University, Shiv Nadar University, DIT University, Dehradun, India; E-mail: ab414@snu.edu.in*

Bhatt, Satvik, *Department of Mechanical Engineering, MNIT Jaipur, India*

Bisht, Dinesh C. S., *Department of Mathematics, Jaypee Institute of Information Technology, Noida - 201304, India; E-mail: drbisht.math@gmail.com*

Chelidze, Tamaz, *Ivane Javakhishvili Tbilisi State University, M. Nodia Institute of Geophysics, Tbilisi, Georgia*

Chitranshi, Mihika, *Department of Mathematics, Lady Shri Ram College for Women, University of Delhi, India; E-mail: mihichi1699@gmail.com*

Dhyani, Bijesh, *Management Studies, Graphic Era Deemed To Be University, Management Studies, Graphic Era University, Graphic Era Hill University, Shiv Nadar University, India; E-mail: bijeshdhyani@gmail.com*

Gopal Naik, M., *Department of Civil Engineering, Osmania University, Hyderabad-500007, India*

Gupta, Harish, *Department of Civil Engineering, Osmania University, Hyderabad-500007, India*

Harini Reddy, M., *Department of Civil Engineering, Annamalai University, Annamalai Nagar-608002, India; E-mail: harinireddy589@gmail.com*

Janiashvili, Manana, *Institute of Clinical Cardiology, 4, Liubliana str. Tbilisi, Georgia*

Joshi, Rahul, *Department of Mechanical Engineering, MNIT Jaipur, India*

Karagrigoriou, A., *University of the Aegean, Greece; E-mail: alex.karagrigoriou@aegean.gr*

Kumar Swamy, N., *School of Sciences, ISBM University Nawapara (Kosmi), Block & Tehsil-Chhura, Gariyaband, Chhattisgarh-493996, India; E-mail: nkumarswamy15@gmail.com*

Kumar, Manish, *Management Studies, Graphic Era Deemed To Be University, Management Studies, Graphic Era University, Graphic Era Hill University, Shiv Nadar University; E-mail: manishsingh12@rediffmail.com*

Kutateladze, Rusudan, *Georgian Technical University, 77, Kostava ave, Tbilisi, Georgia*

Laliashvili, Levan, *Georgian Technical University, 77, Kostava ave, Tbilisi, Georgia; M. Nodia Institute of Geophysics, 1, Alexidze str. Tbilisi, Georgia; Ivane Javakhishvili Tbilisi State University, M. Nodia Institute of Geophysics, Tbilisi, Georgia*

Makris, K. N., *National Technical University of Athens, Greece; E-mail: constantinosmakris@yahoo.gr*

Manikumari, N., *Professor, Department of Civil Engineering, Annamalai University, Annamalai Nagar-608002, India; E-mail: nmanikumariau@gmail.com*

Matcharashvili, Tamar, *Georgian Technical University, 77, Kostava ave, Tbilisi, Georgia*

Matcharashvili, Teimuraz, *Georgian Technical University, 77, Kostava ave, Tbilisi, Georgia; M. Nodia Institute of Geophysics, 1, Alexidze str. Tbilisi, Georgia; Ilia State University, 3/5, Cholokashvili ave. Tbilisi, Georgia; Ivane Javakhishvili Tbilisi State University, M. Nodia Institute of Geophysics, Tbilisi, Georgia*

Mepharidze, Ekaterine, *Ivane Javakhishvili Tbilisi State University, M. Nodia Institute of Geophysics, Tbilisi, Georgia*

Mishra, Nitin, *Assistant Professor, Department of Civil Engineering, Graphic Era Deemed to be University, Dehradun, India; E-mail: nitinuag@gmail.com*

Mohan Raju, M., *Assistant Executive Engineer, Nagarjunasagar Project, Irrigation & CAD (Projects Wing) Department, Govt. of Telangana State, Hill Colony-508202, India; E-mail: mmraju.swce@gmail.com*

Naresh, A., *Department of Civil Engineering, Osmania University, Hyderabad-500007, India; E-mail: ayyaure@gmail.com*

Patil, Amitkumar, *Department of Mechanical Engineering, MNIT Jaipur, India*

Sahu, Sunil K., *School of Sciences, ISBM University Nawapara (Kosmi), Block & Tehsil-Chhura, Gariyaband, Chhattisgarh-493996, India*

Sborshchikovi, Aleksandre, *Ivane Javakhishvili Tbilisi State University, M. Nodia Institute of Geophysics, Tbilisi, Georgia; E-mail: a.sborshchikov@gmail.com*

Singh, Gaurav, *Punjab Remote Sensing Centre, Ludhiana, India; E-mail: gaurav.panwar.gs@gmail.com*

Singh, Yograj, *Department of Mathematics, Lady Shri Ram College for Women, University of Delhi, India; E-mail: yograjchauhan26@gmail.com*

Soni, Gunjan, *Department of Mechanical Engineering, MNIT Jaipur, India*

Tepnadze, Dimitri, *Ivane Javakhishvili Tbilisi State University, M. Nodia Institute of Geophysics, Tbilisi, Georgia*

Thakural, L. N., *Water Resources Division, National Institute of Hydrology, Roorkee, India; E-mail: thakuralln@gmail.com*

Tsveraidze, Zurab, *Georgian Technical University, 77, Kostava ave, Tbilisi, Georgia*

Verma, Poonam, *Management Studies, Graphic Era Deemed To Be University, Management Studies, Graphic Era University, Graphic Era Hill University, Shiv Nadar University, India; E-mail: Poonaddn@gmail.com*

Vonta, I., *National Technical University of Athens, Greece; E-mail: ilia.vonta@math.ntua.gr*

Zhukova, Natalia, *Ivane Javakhishvili Tbilisi State University, M. Nodia Institute of Geophysics, Tbilisi, Georgia*

List of Abbreviations

ACF	Auto-correlation function
ADF	Augmented Dicky Fuller
AHP	Analytic Hierarchy Process
AIC	Akaike's Information Criterion
AMaxT	anomalies of max daily temperatures
AMinT	anomalies of min daily temperatures
ANN	Artificial Neural Network
AR	autoregressive (AR)
ARIMA	auto-regressive integrated moving average
ARMA	autoregressive moving-average
AS	autonomous system
BGP	Border Gateway Protocol
BHHJ divergence	Basu, Harris, Hjort and Jones divergence
BPA	back propagation algorithm
BSE	Bombay Stock Exchange
CBM	Conditional Based Maintenance
CBP	Cascaded back propagation
CC	correlation coefficient
CCCMA	Canadian Centre for Climate Modelling & Analysis
CCIS	Canadian Climate Impacts Scenarios
CE	coefficient of efficiency
C-MAPSS	Commercial Modular Aero-Propulsion Simulation System
C-MAPSS	Commercial Modular Aero-Propulsion Simulation System
CNN	Convolutional Neural Network
CV	Coefficient of Variation
CWC	Central Water Commission
DB	doing business
DC	coefficient of determination

DM	Decision Maker
DWT	discrete wavelet transforms
GARCH	Generalized Auto Regressive Conditional Heteroskedasticity
GCM	General circulation models
GPU	Graphics Processing Unit
HPC	High-Pressure Compressor
HPP	Hydro Power Plant
HPT	High-Pressure Turbine
ICT	Information and Communications Technology
IED	inter earthquake distances
IEF	Index of Economic Freedom
IET	inter earthquake times
ILI	Influenza Like Illness
ILSVRC	ImageNet Large Scale Visual Recognition Challenge
IMD	Indian Meteorological Department
KLD	Kullback-Leibler divergence
LPC	Low-Pressure Compressor
LPT	Low-Pressure Turbine
LRN	Layer Recurrent Network
LSTM	long short-term memory
LZC	Lempel-Ziv Complexity Measure
MA	moving-average
MADM	multiple attribute decision making
MAE	Mean Absolute Error
MAPE	mean absolute percentage error
MAUT	Multiple Attribute Utility Theory
MCDM	Multiple-criteria decision-making
MD	Euclidean distance
ME	Mean Error
MFDFA	Multifractal Detrended Fluctuation Analysis
MLR	multiple linear regression
MODM	Multiple Objective Decision Making
MPE	Mean Percentage Error
MRO	Maintenance Repair and Overhaul
MSE	mean square error
MSE	mean squared error
NCEP	National Centers for Environmental Prediction
NIH	National Institute of Hydrology

NSE	National Stock Exchange
PACF	Partial Autocorrelation Function
PCA	Principal Component Analysis
PHM	Prognostic Health Management
PoF	Physics of Failure
ReLU	Rectified Linear Unit
RMS	root mean square
RMSE	root mean square error
RMSPE	Root Mean Squared Percentage Error
RUL	Remaining Useful Life
SDSM	Statistical Downscaling Model
SES	simple exponential smoothing
SGDM	Stochastic Gradient Descent with Momentum
SRC	Sediment rating curve
SVM	Support Vector Machine
TOPSIS	Technique for Order of Preference by Similarity to Ideal Solution
TPU	Tensor Processing Unit
VGG	Visual Geometry Group
VIKOR	VlseKriterijumska Optimizacija I Kompromisno Resenje

1

On Dimensionless Dissimilarity Measures for Time Series

K. N. Makris[1], A. Karagrigoriou[2], and I. Vonta[3]

[1]National Technical University of Athens, Greece
[2]University of the Aegean, Greece
[3]National Technical University of Athens, Greece
E-mail: constantinosmakris@yahoo.gr, alex.karagrigoriou@aegean.gr, ilia.vonta@math.ntua.gr

Abstract

Dissimilarity measures have a pivotal role in applied and computational statistical data analysis. This chapter deals with an overview of the topic of dissimilarity measures and introduces a special class of dimensionless dissimilarity measures associated with the coefficient of variation with applications in time series analysis.

Keywords: divergence measure, coefficient of variation, similarity measures, time series, dimensionless indices.

1.1 Introduction

An issue of fundamental importance in Statistics is the comparison of two distributions, two populations or in general, two functions. In practice it is often needed to decide whether two populations have or not the same distribution. For such purposes we rely on quantities known as divergence or dissimilarity measures. Through the dissimilarity measures we intend to quantify and compare the information or the uncertainty associated with random variables. The classical measures are classified in four classes namely divergence - type, entropy - type, Fisher - type and Bayesian - type measures (see Vonta and Karagrigoriou, 2011).

The classical Pearson's chi-square test can be considered as one of the first dissimilarity measures introduced by Pearson (1900). At the same time, the concept of entropy has a long history dated back to the 19th century. More specifically, Boltzmann discussed the so called second law of thermodynamics and laid the foundation of the statistical approach to thermodynamics. Boltzmann's entropy which was rigorously established by Planck in 1900, is the basis to all statistical concepts of entropy.

Shannon (1948) suggests and examines the notions of entropy and mutual information. The mutual information measures the mutual dependence between two variables by quantifying the "amount of information" collected regarding one of the variables by observing the other.

The relation between Information Theory and Statistics was proposed by Kullback and Leibler (1951) who extended the notion of Shannon's entropy and proposed a measure known as Kullback-Leibler Divergence or Relative Entropy. Their work on *Information Theory and Statistics* initiated the research on the field of Statistical Information Theory. Before Kullback and Leibler, many scientists (e.g. Mahalanobis (1936); Bhattacharyya (1943)) proposed various types of dissimilarity but it was the work by Kullback and Leibler that made the divergence measures mainstream to the scientific community. Dissimilarity measures have numerous applications in many often diverse fields including Applied Mathematical and Physical Sciences, Probability Theory, Stochastic Processes, Statistics and Financial Mathematics.

Cressie and Read (1984) attempted to provide a unified analysis by introducing the so called power divergence family of statistics that involved an index and is used in tests of fit. The Cressie and Read family includes a number of well known measures like the Pearson's dissimilarity measure or the standard log-likelihood ratio statistic, G^2. In 1998 the BHHJ divergence measure was introduced (Basu et al., 1998) as a robust estimating procedure which later formed the basis of a new general family of dissimilarity measures (Mattheou et al., 2009) for hypothesis testing purposes. The BHHJ family, like the Cressie and Read family, relies on an index which controls the trade-off between robustness and efficiency.

It is well known that dissimilarity measures have a central role in applied statistics with numerous applications. They can be used for inferential statistics like estimating purposes (Toma 2008; 2009) and hypothesis testing problems (e.g., Huber-Carol et al., 2002; Zhang, 2002; Meselidis and Karagrigoriou, 2020). Furthermore, it should be pointed out that the relative entropy has been used for the development of model identification procedures like the well-known Akaike Information Criterion (e.g. Karagrigoriou, 1995; Cavanaugh, 2004; Shang, 2008; Mantalos et al., 2010). Applications of dissimilarity measures can also be found in biostatistics, survival analysis or reliability theory where the concept of censoring which constitutes a challengng matter, is often present.

The dissimilarity on time series is a problem with one extra special characteristic, that of dependence accompanied by the concept of (time) ordering. The problem arises when two (or more) time series are compared to establish whether they are sharing a common behavior (distribution). For such problems one may rely on classical or advanced mathematical metrics, on dissimilarity measures, on data transformations or on algorithmic approaches.

In this chapter we discuss and examine a series of dissimilarity measures including a discussion about their capabilities and their advantages and disadvantages. One of the main contributions of the present work is the proposal of an advanced dimensionless dissimilarity measure which takes advantage of the first two moments of the distribution to measure the dissimilarity between two series. Sections 1.2 and 1.3 are devoted to standard mathematical and divergence measures. Section 1.4 deals with a class of classical and advanced dissimilarity measures for time series data including the dimensionless dissimilarity measure based on the coefficient of variation.

1.2 Classical Dissimilarity Measures

The dissimilarity measures are used for the quantification of the distance or closeness of functions, populations or data sets. The standard property required to be satisfied by a dissimilarity measure is the nonnegativity. Other properties include but not necessarily, the symmetry and the triangular inequality. If a dissimilarity measure does satisfy all the above properties then it is considered to be a classic metric. It should be noted that for statistical purposes the nonnegativity is the only required property. Well-known measures, like for example the relative entropy, are not symmetric which though, in certain instances, is a serious defect.

Consider the random variables X and Y with realizations $\mathbf{x} = (x_1, \ldots, x_k)'$ and $\mathbf{y} = (y_1, \ldots, y_k)'$. Then, representative examples of classical measures include, among others, the Euclidean distance given by

$$d(X, Y) = ||X - Y|| = \sqrt{\sum_{i=1}^{k}(x_i - y_i)^2} = \sqrt{(X - Y)'(X - Y)}.$$

and the City-Block distance defined by

$$d_C(X, Y) = \sum_{i=1}^{k} |x_i - y_i|.$$

It should be noted that most of the classical measures do not take into consideration variances or covariances among the variables involved. The Mahalanobis distance comes to fill in the gap by incorporating into the measure the variance-covariance matrix:

$$D_{Mh}(X, Y) = (X - Y)' \Sigma^{-1} (X - Y)$$

where Σ^{-1} is the inverse of the variance-covariance matrix of X and Y.

1.3 Classical Entropy-type Dissimilarity Measures

In this section we present the classical dissimilarity measures associated with the concept of entropy which was initially defined and used in Physics, in the field of thermodynamics with its formal probabilistic definition being attributed to Boltzmann (1866).

One of the classical entropic dissimilarity measures is the relative entropy between two densities $f(x)$ and $g(x)$ for the random variable X which is defined by

$$I_X^{KL}(f, g) = \int f \ln(f/g) d\mu = E_f \left[\ln(f/g) \right]. \tag{1.1}$$

The discrete equivalent of the above dissimilarity measure for two distributions P and Q with probability mass functions $p = (p_1, \ldots, p_k)'$ and $q = (q_1, \ldots, q_k)'$ for the random variable X, is given by

$$I_X^{DKL}(P, Q) = \sum_{i=1}^{k} p_i \ln(p_i/q_i) = E_P \left[\ln(P/Q) \right]. \tag{1.2}$$

The above measure is often used to compare a theoretical distribution f or P with a candidate model. In such cases various g and Q candidate models could be used and the one for which the relative entropy is the smallest is chosen as a good estimate of the theoretical distribution f or P. In that sense the relative entropy can be considered as a model selection or identification technique.

It is easy to observe that the relative entropy consists of two terms. Indeed, for instance for the discrete case we have

$$I_X^{DKL}(P,Q) = \sum_{i=1}^{k} p_i \ln(p_i) - \sum_{i=1}^{k} p_i \ln(q_i). \qquad (1.3)$$

Observe that for the model selection approach mentioned above the first term of the relative entropy is fixed while the second changes according to the candidate model used. As a result one could focus solely on the second term which is known as cross entropy given by

$$I_X^{CE}(P,Q) = -\sum_{i=1}^{k} p_i \ln(q_i). \qquad (1.4)$$

The relative entropy which is also known as Kullback-Liebler dissimilarity measure (Kullback and Liebler, 1951) is not symmetric, a disadvantage resolved by the so called Jeffrey's dissimilarity measure defined below (Jeffreys, 1946):

$$I_X^{JRE}(P,Q) = \sum_{i=1}^{k} p_i \ln(p_i/q_i) + \sum_{i=1}^{k} q_i \ln(q_i/p_i). \qquad (1.5)$$

It should be mentioned that all the dissimilarity measures mentioned above have their origin in Shannon's entropy (Shannon, 1948) given by

$$I^S(X) \equiv I^S(f) = -\int f \ln f \, d\mu = E_f[-\ln f],$$

where X is a random variable with density function $f(x)$ and μ a probability measure on $\mathbb{R}$.

For generalizations and extensions of Shannon's entropy the interested reader is referred among others, to the work of Rényi (1961), Liese and Vajda (1987) Nadarajah and Zografos (2003) and Zografos and Nadarajah (2005). As for extensions of the relative entropy we should mention the work of Csiszar (1963), Cressie and Read (1984), Basu et al. (1998), Mattheou et al. (2009), Toma (2009), and Toma & Broniatowski (2011).

1.4 Dissimilarity Measures for Time Series Data

Let us consider two time series of equal length $\{X_t^1\}_{t=1}^{T}$ and $\{X_t^2\}_{t=1}^{T}$ and assume that for $j = 1, 2$ and $i = 1, \ldots, T$

- $X^j_{(i)}$ represents the ith ordered (ranked) observation of the jth time series and
- T^j_i is equal to the time point t, $t \in \{1, 2, \ldots, T\}$ at which the ith ordered observation of the jth time series has been observed.

Hence, T^j_i is the time point t of the $X^j_{(i)}$ ordered (ranked) observation. Finally, $K(M)$ represents the number of consecutive extreme values (either highest if $M = 1$ or lowest if $M = 0$) used for the analysis with $K = 1, \ldots, T$. For instance, $K(M) = 4(1)$ refers to 4 *highest* values of a series $\{X_t\}_{t=1}^T$ denoted by $X_{(T)}, X_{(T-1)}, X_{(T-2)}$ and $X_{(T-3)}$ while $K(M) = 4(0)$ refers to the 4 *lowest* values denoted by $X_{(1)}, X_{(2)}, X_{(3)}$ and $X_{(4)}$. Naturally if $K = T$ then the entire data set is used and the index M becomes redundant.

The purpose of dissimilarity is to measure the degree (extend) of similarity of two time series. The measures to be discussed in the next two subsections are examining the closeness of the K extreme values of such ordered data. In addition to standard measures defined in the first subsection, the second is devoted to the proposal of a dimensionless dissimilarity measure based on the coefficient of variation. If they coincide, the dissimilarity will be equal to 0 indicating that the data sets are identical. The bigger the difference, the higher the dissimilarity.

1.4.1 Standard Dissimilarity Measures

Consider two time series $\{X^1_t\}_{t=1}^T$ and $\{X^2_t\}_{t=1}^T$ where $X^1_{(i)}$ and $X^2_{(i)}$ are the corresponding ordered observations. Let also T^1_i and T^2_i represent the time points of the $X^1_{(i)}$ and $X^2_{(i)}$ ordered series. The first definition refers to a measure that compares the extreme values of a series to the observations of another series in terms of occurrence at the exact same time points.

Definition 1. The $K - similarity$ measure between two time series $\{X^1_t\}_{t=1}^T$ and $\{X^2_t\}_{t=1}^T$ is defined by

$$V_{X^1, X^2}[K(M)] = \begin{cases} \sum_{i=1}^K \left(X^1_{(i)} - X^2_{T^1_i} \right)^2 & \text{if } M = 0 \\ \sum_{i=1}^K \left(X^1_{(T-i+1)} - X^2_{T^1_{T-i+1}} \right)^2 & \text{if } M = 1 \end{cases} \quad (1.6)$$

In the above definition observe that we first identify the K highest (if $M = 1$) or lowest observations (if $M = 0$) of the first time series together with the corresponding time points T^1_i in which they have been observed. Then, these

1.4 Dissimilarity Measures for Time Series Data 7

K values are compared one by one with the corresponding observations of the second time series observed at the exact same time points.

If instead we are interested in examining whether the shapes of the series are similar without being of the same magnitude we could concentrate on the time points T_i^j of occurrence of the K extreme observations. The relevant definition follows.

Definition 2. Let T_i^j represent the time point of the $X_{(i)}^j$ ordered observation, $i = 1, \ldots, T$, $j = 1, 2$. The K time–similarity measure between two time series $\{X_t^1\}_{t=1}^T$ and $\{X_t^2\}_{t=1}^T$ is defined by

$$V_{1,2}[K(M)] = \begin{cases} \sum_{i=1}^{K} \left(T_i^1 - T_i^2\right)^2 & \text{if } M = 0 \\ \sum_{i=1}^{K} \left(T_{T-i+1}^1 - T_{T-i+1}^2\right)^2 & \text{if } M = 1 \end{cases} \quad (1.7)$$

The shape of a series is often important in medical, financial or meteorological and environmental data where a seasonality pattern often is present. As a result the importance of the above definition is evident where two series are similar if K extreme observations occurred at the same time points irrespectively of their magnitude. Thus, if the researcher investigates the repetition in time in time series with seasonality he/she should choose to apply Definition 2 instead of Definition 1. It goes without saying that the measure can be evaluated for any K, $K = 1, 2, \ldots, T$.

In cases such as those mentioned above it is not uncommon to observe a similar seasonal pattern shifted or delayed in time. Indeed, in such cases the occurrence of the same pattern may occur with a time delay between seasons. For such situations the following definition applies:

Definition 3. Let T_i^j represent the time point of the $X_{(i)}^j$ ordered observation, $i = 1, \ldots, T$, $j = 1, 2$ and m be any integer. The K (time-shift)–similarity measure between two time series $\{X_t^1\}_{t=1}^T$ and $\{X_t^2\}_{t=1}^T$ is defined by

$$V_{1,2}^m[K(M)] = \begin{cases} \sum_{i=1}^{K} \left(T_i^1 - T_{i+m}^2\right)^2 & \text{if } M = 0 \\ \sum_{i=1}^{K} \left(T_{T-i+1}^1 - T_{T-i+1-m}^2\right)^2 & \text{if } M = 1. \end{cases} \quad (1.8)$$

Note that for $m = 0$ the Definition reduces to Definition 2.

The following two simple examples unfold the basic characteristics of the measures defined earlier.

8 *On Dimensionless Dissimilarity Measures for Time Series*

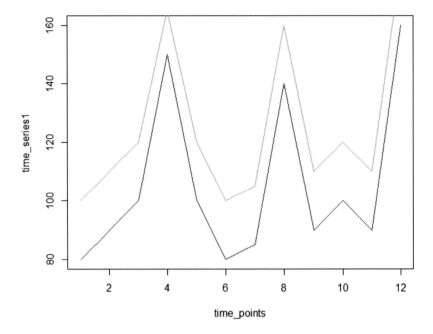

Figure 1.1 Two time series, n=12 (series 1 in red and series 2 in green).

Example 1. It is easily observed from Figure 1.1 that the two time series have the same shape, namely their ordered observations occurred at the exact same time points. As a result the measure $V_{1,2}^m[K(M)]$ for $M = m = 0$ is equal to 0 for any value of $K \in \{1, \ldots, T\}$. At the same time though the observations do not have the same magnitude and in fact the measure $V_{X^1,X^2}[K(M)]$ is significantly larger that 0 for any K. For instance, the observations differ by 20 units for each of the five lower ordered observations so that $V_{X^1,X^2}[5(0)] = 5 * 20^2$. This is a representative example where the two series are similar in terms of the shape but not in terms of the magnitude.

Example 2. In Figure 1.2 consider the 3 higher ordered observations (i.e. $K = 3$ and $M = 1$). It is easily seen that although the maximum observation on each data set occurred at the time point $T_{12}^1 = T_{12}^2 = 12$ and therefore $V_{1,2}^m[1(1)] = 0$ and $V_{X^1,X^2}[1(1)] = 20^2$ the same is not true for other values. Thus, $T_{11}^1 = 4$, $T_{11}^2 = 3$ and $T_{10}^1 = 8$ and $T_{10}^2 = 5$ so that $V_{1,2}^m[2(1)] = 1$ and $V_{1,2}^m[3(1)] = 10$. Note that for ranking ties, the chronological order is taken into consideration with the rank of the first occurence being the largest and the rank of the last occurence being the smallest.

1.4 Dissimilarity Measures for Time Series Data   9

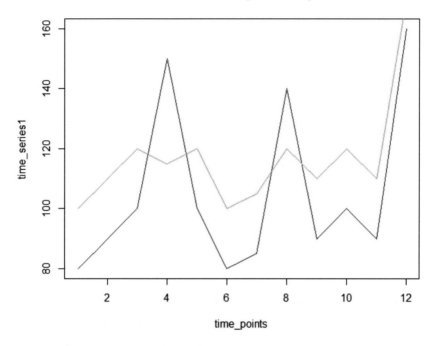

Figure 1.2 Two time series, n=12 (series 1 in red and series 2 in green).

It is easily observed that if the measure in Equation (1.6) is equal to zero that means that the two series are similar in the sense that they have the exact same K values occurring on the exact same time points. It should be noted though that although for the X^1 series the values are the K extreme values for the X^2 series the same values may not necessarily be the K extreme values. A way to resolve this issue is to reevaluate the measure in Equation (1.6) by interchanging the roles of the two series involved and combine the resulting measures into a single symmetric measure.

Definition 4. The $K-(s)similarity$ measure between two time series $\{X^1_t\}_{t=1}^T$ and $\{X^2_t\}_{t=1}^T$ with (s) standing for "symmetric", is denoted by $V^S_{X^1,X^2}[K(M)]$ and is defined by

$$V^s_{X^1,X^2}[K(M)] = \begin{cases} \sum_{i=1}^{K}\left\{\left(X^1_{(i)}-X^2_{T^1_i}\right)^2+\left(X^2_{(i)}-X^1_{T^2_i}\right)^2\right\} & \text{if } M=0 \\ \sum_{i=1}^{K}\left\{\left(X^1_{(T-i+1)}-X^2_{T^1_{T-i+1}}\right)^2+\left(X^2_{(T-i+1)}-X^1_{T^2_{T-i+1}}\right)^2\right\} & \text{if } M=1 \end{cases}$$
(1.9)

An alternative way to handle the issue is by comparing the two $K-$ similarity measures based on X^1 and X^2. The resulting measure is defined below:

Definition 5. The $K-(rs)similarity$ measure between two time series $\{X_t^1\}_{t=1}^T$ and $\{X_t^2\}_{t=1}^T$ with (rs) standing for "ratio summation", is denoted by $V_{X^1,X^2}^{rs}[K(M)]$ and is defined by

$$V_{X^1,X^2}^{rs}[K(M)] = \begin{cases} \dfrac{\sum_{i=1}^{K}\left(X_{(i)}^1 - X_{T_i^1}^2\right)^2}{\sum_{i=1}^{K}\left(X_{(i)}^2 - X_{T_i^2}^1\right)^2} & \text{if } M = 0 \\[2em] \dfrac{\sum_{i=1}^{K}\left(X_{(T-i+1)}^1 - X_{T_{T-i+1}^1}^2\right)^2}{\sum_{i=1}^{K}\left(X_{(T-i+1)}^2 - X_{T_{T-i+1}^2}^1\right)^2} & \text{if } M = 1. \end{cases} \quad (1.10)$$

Note that the notion of comparison could be used for the time points T_i^j in a fashion similar to the one presented in Definition 5. Indeed, the following definition clarifies the issue.

Definition 6. Let T_i^j represent the time point of the $X_{(i)}^j$ ordered observation, $i = 1, \ldots, T$, $j = 1, 2$. The $K-(rs)time-similarity$ measure between two time series $\{X_t^1\}_{t=1}^T$ and $\{X_t^2\}_{t=1}^T$ is denoted by $V_{1,2}^{rs}[K(M)]$ and is defined by

$$V_{1,2}^{rs}[K(M)] = \frac{\sum_{i=1}^K T_i^1}{\sum_{i=1}^K T_i^2}. \quad (1.11)$$

Remark 1. Observe that a definition similar to Definition 5 or Definition 6 could be based on the product instead of the summation with a similar effect. The resulting measures will be called $K-(r\prod)-similarity$ and $K-(r\prod)time-similarity$.

The following two simple examples present the special features associated with the measures defined above.

Example 3. Observe that the above two special measures may produce misleading results because it is possible that the K extreme times may be the same without being in one-to-one correspondence. Indeed, assume that the 5 highest observed daily values (i.e. $K = 5$, $M = 1$) of two series of length

$n = 40$ occur at the same time points, say $T_{36} - T_{40}$ but not all with the same order. For instance $T^1_{36} = T^2_{40}$, say equal to 30 (30th day) and $T^1_{40} = T^2_{36}$, say equal to 20 (20th day) with the other three being identical. Hence, naturally $V^S_{1,2}[K(M)] = V^\Pi_{1,2}[K(M)] = 1$. On the other hand, $V_{1,2}[K(M)] = 200$.

Example 4. Consider another example where all but one of the 5 highest observations occur at exactly the same time points, with just one occurring on extremely different time points, say, for instance that, $T^1_{36} = 5$ and $T^2_{36} = 40$. In such a case the Euclidean type measure $V_{1,2}[K(M)]$ will be quite large ($35^2 = 1225$) while the summation type measure will be equal to $(40 + x)/(5 + x) < 8$ irrespectively of the value of x. Similarly for the product measure.

The following Lemma provides the basic properties for the proposed measures which are straightforward:

Lemma 1. Consider two time series $\{X^1_t\}^T_{t=1}$ and $\{X^2_t\}^T_{t=1}$. Let T^j_i represent the time point of the $X^j_{(i)}$ ordered observation, $i = 1, \ldots, T$, $j = 1, 2$. Then, for the measures defined in this section the following properties hold true:

1. $V_{j,j}[K(M)] = \frac{1}{K}\sum_{i=1}^{K}\left(T^j_i - T^j_i\right)^2 = 0$
2. $V_{X^j, X^j}[K(M)] = 0$, $j = 1, 2$
3. $V_{1,2}[K(M)] = V_{2,1}[K(M)]$
4. $V^s_{X^1, X^2}[K(M)] = V^s_{X^2, X^1}[K(M)]$
5. $V_{1,2}[K(M)] \geq 0$
6. $V^s_{X^1, X^2}[K(M)] \geq 0$
7. $V^{rs}_{1,2}[K(M)] \geq 0$.

Remark 2. From the above properties we easily deduce that the $K-$similarity measure given in Equation (1.6) does not satisfy the symmetric property which implies that although it is a distance measure it is not a proper metric.

Remark 3. It is also possible to combine the time points together with the actual observations (e.g Equations (1.10) and (1.11)) to make it possible to distinguish between series that are similar for just one of the two measures. In such cases we could also introduce weights on the two components of the combined measure. In such a case we could define the *weighted − combined − K − measure* as follows:

Definition 7. Let T^j_i represent the time point of the $X^j_{(i)}$ ordered observation, $i = 1, \ldots, T$, $j = 1, 2$. Let also $w_1, w_2 \geq 0$ with $w_1 + w_2 = 1$.

The $weighted - combined - K - measure$ is the combination of the $K - (s)similarity$ measure and the $K\ time - similarity$ between two time series $\{X_t^1\}_{t=1}^T$ and $\{X_t^2\}_{t=1}^T$ is denoted by $V^w[K(M)]$ and is defined by

$$V^w[K(M)] = w_1 V_{X^1,X^2}^s[K(M)] + w_2 V_{1,2}[K(M)]. \qquad (1.12)$$

Observe that the original individual dissimilarity measures are special cases of the above combined measure for $w_1 = 1, w_2 = 0$ and $w_1 = 0, w_2 = 1$, respectively.

Closing this section we conclude that the previously defined measures could be in some instances, unstable depending on the special features of the K extreme values of the time series involved. In the next subsection we will attempt to handle such matters by proposing stable dimensionless dissimilarity measures without the defects of the above measures.

1.4.2 Advanced Dissimilarity Measures

In this section we present a new dimensionless measure which attempts to overcome the defects associated with standard measures presented in the previous subsection. In Makris et al. (2021) the so called $K-dimensionless-coefficient$ which is based on the average among either the K observed ordered values or the corresponding time points where these values occur was introduced. These measures were introduced in Makris (2017) for experimental data for two wind turbines and later by the same author (Makris, 2018) for incidence rate data of influenza like illness (ILI). Such measures although useful in several instances they are heavily affected by extreme observations and at the same time they are taking into consideration only the first moment without taking into consideration the variability associated with the data set under investigation. In this section we generalize this definition by considering the coefficient of variation (CV) which takes into account both the mean and the standard deviation of a series. The definition below provides the $K - dimensionless - coefficient$ of Makris et al. (2021).

Definition 8 (Makris et. al, 2021). Let $X_{(i)}^j$, $i = 1, 2, \ldots, n$ be the ordered observations of the j series, $j = 1, 2$. Let also T_i^j be the time point associated with the ith ordered observation of the jth series. Then, the $K - dimensionless - coefficient$ for a series X is denoted by $D_X[KM]$ and is defined by

$$D_X[KM] = \begin{cases} \dfrac{Average\{|X_{(i)}|, i=n-K+1,\ldots,n\}}{Average\{|X_{(i)}|, i=1,\ldots,n\}}, & \text{if } M = 1 \\[2ex] \dfrac{Average\{|X_{(i)}|, i=1,\ldots,K\}}{Average\{|X_{(i)}|, i=1,\ldots,n\}}, & \text{if } M = 0. \end{cases} \quad (1.13)$$

The definition below is the proposed $K-dimensionless-CV-measure$ based on the coefficient of variation and in fact it compares the CV of the entire series with the CV of the portion of the series which is of interest to the researcher. This definition provides significant information about the variability around the mean of the entire series as compared with a specific portion of interest.

Definition 9. Let $X^1_{(i)}, i = 1, 2, \ldots, n$ be the ordered observations of a series. Let also T^1_i be the time point associated with the ith ordered observation of the series. Then, the $K-dimensionless-CV-measure$ for a series X^1 is defined by

$$DCV_{X^1}[KM] = \begin{cases} \dfrac{CV\{X^1_{(i)}, i=n-K+1,\ldots,n\}}{CV\{X^1_{(i)}, i=1,\ldots,n\}}, & \text{if } M = 1 \\[2ex] \dfrac{CV\{X^1_{(i)}, i=1,\ldots,K\}}{CV\{X^1_{(i)}, i=1,\ldots,n\}}, & \text{if } M = 0. \end{cases} \quad (1.14)$$

As expected for $K = T$ the entire series is used and the coefficient is equal to 1. It should be noted that the proposed measure relies on the original observations even if they are negative in order to be used for comparative purposes against other series. Thus, the following definition provides the comparative coefficient between two series X and Y.

Definition 10. Let $X^j_{(i)}, i = 1, 2, \ldots, n$ be the ordered observations of the j series, $j = 1, 2$. Let also T^j_i be the time point associated with the ith ordered observation of the jth series. Then, the $K-dimensionless-CV-measure$ between the two series is defined by

$$DCV_{X^1, X^2}[KM] = \dfrac{DCV_{X^1}[KM]}{DCV_{X^2}[KM]} \quad (1.15)$$

It is worth mentioning that the similarity of two series based on the DCV measure implies that the coefficient of variation of a part of the distribution as compared with the CV of the entire data set is similar to the corresponding comparative CV of another series. The advantage of the proposed measure is that it takes into consideration the first two moments of the series making the comparison more realistic. It should be noted that the comparative measure could be based only on the portion of the two series without considering the overall CV of the entire series. In such a case, the previous definition simplifies to the definition below:

Definition 11. Let $X_{(i)}^j$, $i = 1, 2, \ldots, n$ be the ordered observations of the j series, $j = 1, 2$. Let also T_i^j be the time point associated with the ith ordered observation of the jth series. Then, the $K-dimensionless-measure$ between the two series is defined by

$$DCV0_{X^1,X^2}[KM] = \begin{cases} \dfrac{CV\{X_{(i)}^1, i=n-K+1,\ldots,n\}}{CV\{X_{(i)}^2, i=n-K+1,\ldots,n\}}, & \text{if } M = 1 \\[2ex] \dfrac{CV\{X_{(i)}^1, i=1,\ldots,K\}}{CV\{X_{(i)}^2, i=1,\ldots,K\}}, & \text{if } M = 0. \end{cases} \quad (1.16)$$

The following Lemma provides the basic properties for the proposed measures which are straightforward:

Lemma 2. Consider two time series $\{X_t^1\}_{t=1}^T$ and $\{X_t^2\}_{t=1}^T$ and Let T_i^j represent the time point of the $X_{(i)}^j$ ordered observation, $i = 1, \ldots, T$, $j = 1, 2$. Then, for the measures defined in this section the following properties hold true:

- $D_X[TM] = 1$ for both $M = 0$ and $M = 1$
- $D_Y[KM] \geq 1$ for $M = 1$ and $Y_i \geq 0$.
- $D_Y[KM] \leq 1$ for $M = 0$ and $Y_i \geq 0$.
- $D_Y[KM] = D_X[KM]$ for a specific K means that the series are similar in the sense of Definition 8
- $D_X[K_1M] \leq D_X[K_2M]$ for $K_1 > K_2$ and $M = 1$
- $D_X[K_1M] \geq D_X[K_2M]$ for $K_1 > K_2$ and $M = 0$
- $DCV0_{X^1,X^2} \geq 1$ implies that the variability of the K ordered observations of the first series around the mean is greater than that of the second series.

1.5 Conclusions

This work deals with measures of dissimilarity with the emphasis being on ordered data such as time series data. It should be noted though that such measures could also be implemented in cases with data characterized by independence. A series of classical dissimilarity measures based both on the actual observations and the time points the observations occurred, have been presented together with a new dimensionless dissimilarity measure that takes into consideration the first two moments of the distribution involved. The proposed measures appear to be both attractive and easy to be implemented with great applicability in many diverse fields where ordered data are involved.

References

Basu, A., Harris, I. R., Hjort, N. L., Jones, M. C. (1998). Robust and efficient estimation by minimising a density power divergence. *Biometrika*, 85, 549–559.

Bhattacharyya, A. (1943). On a measure of divergence between two statistical populations defined by their probability distributions. *Bull. Calcutta Math. Soc.*, 35, 99–109.

Cavanaugh, J. E. (2004). Criteria for linear model selection based on Kullback's symmetric divergence. *Aust. N. Z. J. Stat.* 46, 257–74.

Cressie, N., Read, T. R. C. (1984). Multinomial goodness-of-fit tests. *J. Roy. Stat. Soc. B*, 5, 440–454.

Csiszar, I. (1963). Eine Informationstheoretische Ungleichung und ihre Anwendung auf den Bewis der Ergodizitat on Markhoffschen Ketten. *Publication of the Mathematical Institute of the Hungarian Academy of Sciences*, 8, 84–108.

Huber-Carol, C., Balakrishnan, N., Nikulin, M. S., and Mesbah, M. (2002). *Goodness-of-fit Tests and Model Validity*, Birkhäuser, Boston.

Jeffreys, H. (1946). An invariant form for the prior probability in estimation problems. *Proceedings of the Royal Society of London. Series A. Mathematical and Physical Sciences*, 186(1007), 453–461.

Karagrigoriou, A. (1995). Asymptotic efficiency of model selection criteria: The non-zero mean Gaussian AR infinity case. *Comm. Statist. Theory & Meth*, 24, 911–930.

Kullback, S., Leibler, R. (1951). On information and sufficiency. *Ann. Math. Stat.*, 22, 79–86.

Liese, F. and Vajda, I. (1987). *Convex Statistical Distances*. Teubner, Leipzig.

Mahalanobis, P. C. 1936). On the generalized distance in statistics. *National Institute of Science of India*, 2, 49–55.

Makris, K. (2017). Statistical Analysis of random waves in SPAR-type and TLP-type wind turbines, *MSc. Thesis*, National Technical University of Athens, Greece (in Greek).

Makris, K. (2018). Statistical Analysis of epidemiological time series data, *MSc. Thesis*, National Technical University of Athens, Greece (in Greek).

Makris, K., Karagrigoriou, A., Vonta, I. (2021). On divergence and dissimilarity measures for multiple time series, In *Applied Modelling Techniques and Data Analysis*, Dimotikalis, I. et al. iSTE Wiley, 249–261.

Mattheou, K., Lee, S., and Karagrigoriou, A. (2009). A model selection criterion based on the BHHJ measure of divergence. *J. Statist. Plann. Infer.*, 139, 128–135.

Mantalos, P., Mattheou, K., Karagrigoriou, A. (2010). An improved divergence information criterion for the determination of the order of an AR process. *Comm. in Statist. – Comput. Simul.*, 39 (5), 865–879.

Meselidis, C. and Karagrigoriou, A. (2020). Statistical inference for multinomial populations based on a double index family of test statistics. *J. of Statist. Comput. and Simul.*, 90(10), 1773–1792.

Nadarajah, S. and Zografos, K., (2003). Formulas for Renyi information and related measures for univariate distributions. *Information Sciences*, 155, 118–119.

Pearson, K. (1900). On the criterion that a given system of deviations from the probable in the case of a correlated system of variables is such that it can be reasonably supposed to have arisen from random sampling. *The London, Edinburgh, and Dublin Philosophical Magazine and Journal of Science*, 50(302), 157–175.

Renyi, A. (1961). On measures of entropy and information. *Proc. of the Fourth Berkeley Symposium on Mathematical Statistics and Probability*, 1, 547–561.

Shang, J. (2008). Selection criteria based on Monte Carlo simulation and cross validation in mixed models. *Far East J. Theor. Statist.*, 25, 51–72.

Shannon, C. E. (1948). A mathematical theory of communication. *Bell System Technical Journal*, 27, 379–423.

Toma, A. (2008). Minimum Hellinger distance estimators for multivariate distributions from the Johnson system. *J. Statist. Plan. and Infer.*, 138, 803–816.

Toma, A. (2009). Optimal robust M-estimators using divergences. *Statistics and Probability Letters*, 79, 1–5.

Toma, A. and Broniatowski, M. (2011). Dual divergence estimators and tests: robustness results. *J. Multivariate Anal.*, 102(1), 20–36.

Vonta, F. and Karagrigoriou, A. (2011). Information measures in biostatistics and reliability, In *Mathematical and Statistical Models and Methods in Reliability*, Rykov, V. V., Balakrishnan, N., and Nikulin (eds), Birkhauser, Boston, 401–413.

Zhang, J. (2002). Powerful goddness-of-fit tests based on likelihood ratio. *J. R. Stat. Soc. Ser. B* 64(2), 281–294.

Zografos, K. and Nadarajah, S. (2005). Survival exponential entropies. *IEEE Trans. Inform. Theory*, 51, 1239–1246.

2

The Classification Analysis of Variability of Time Series of Different Origin

Teimuraz Matcharashvili[1,2,4], Manana Janiashvili[3], Rusudan Kutateladze[1], Tamar Matcharashvili[1], Zurab Tsveraidze[1], and Levan Laliashvili[1,2]

[1]Georgian Technical University, 77, Kostava ave, Tbilisi, Georgia
[2]M. Nodia Institute of Geophysics, 1, Alexidze str. Tbilisi, Georgia
[3]Institute of Clinical Cardiology, 4, Liubliana str. Tbilisi, Georgia
[4]Ilia State University, 3/5, Cholokashvili ave. Tbilisi, Georgia

Abstract

At present, it can be listed many different research fields in which modern methods of linear and nonlinear data analysis are successfully used. These methods help at qualitative and/or quantitative assessment of dynamical properties of processes taking place in complex systems. Generally, these may be processes taking place in both models as well as in real-world systems. Indeed, there can be listed a number of examples in different areas ranging from atmosphere and geophysics to processes in physiology and Core of the Internet, etc. Therefore, it is understandable why theoretical and practical aspects of the analysis of complex time series remain as one of the main subjects of interdisciplinary research interests. Nowadays, there are known different conceptual solutions in the complex time series analysis, though problems arise when available real-world measurement datasets do not fulfill strong requirements of contemporary data analysis. In such cases, having at hand not enough long time series of imperfect quality, researchers usually are forced to combine different approaches in order to have at least some understanding on general features of targeted complex processes. Classification analysis, combined with concepts of contemporary complex data analysis, is deemed to be one of such solutions for relatively short datasets.

Generally, the main objective of classification or clustering algorithms is to separate classes (or clusters). Different methods to achieve this task have been developed up to date. The most practical and often used way of clustering or classification is the use of methods to find centres that represent clusters and assess their similarity by measures used in the distance metrics, such as Euclidean distance (MD) or Kullback-Leibler divergence (KLD).

Mahalanobis distance calculation combined with the surrogate time series testing is already used for complex time series analysis. In this research, in order to assess the similarity (or dissimilarity) of considered processes (i.e. corresponding time series), we also used the KLD as a similarity measure. Thus, MD and KLD have been used to assess the similarity of dynamics in case of model as well as real-world time series. Exactly, short model datasets generated by low-dimensional attractors, Lorenz and Henon, with and without the added white noises have been analysed. Next, five types of real-world datasets have also been analysed. Namely: (1) Border Gateway Protocol (BGP) time series from four autonomous systems (AS), (2) datasets of inter earthquake times (IET), as well as inter earthquake distances (IED), obtained from the seismic catalogue of southern California, (3) datasets of the yearly number of days when anomalies of max daily temperatures (AMaxT) were significantly larger or were significantly lower (AMinT) than mean values calculated for the analysed period in Tbilisi, Georgia, (4) arterial systolic and diastolic blood pressure, as well as heart rate variability time series of healthy persons and patients with arterial hypertension collected at Institute of Clinical Cardiology, Tbilisi, Georgia and (5) components of the Index of Economic Freedom and Doing Business index datasets for three southern Caucasian countries, obtained from corresponding international databases.

2.1 Introduction

Analysing dynamical features of real-world processes, in most cases, we deal with relatively short and an imperfect quality time series. On the other hand, usually, most standard approaches in modern data analysis require long time series of high quality. These requirements usually are difficult to be fulfilled for the real-world measurement time series. Among many other attempts to resolve this problem in the last decade, we observe efforts aimed at the combination of concepts of the modern complex data analysis with the well-known statistical views of similarity measures. One of such, useful for short time series, approach is based on Mahalanobis distance calculation in combination with the surrogate time series testing (Matcharashvili et al.,

2017). The well-known MD metric is important for classification purposes as it incorporates information about the spatial distribution of points. The similarity or dissimilarity of compared groups is assessed by a number that measures how distinct datasets are. The greater the number, the more distinct the datasets (compared groups) are. One of the other similarity measures is the Kullback–Leibler divergence (KLD). KLD is a measure based on the relative entropy of two probability density functions build for considered processes.

In this research, we have compared the effectiveness of MD and KLD calculation in the classification (or separation) task comparing two dynamical processes (given by the available time series). These have been time series with already detected earlier, internal nonlinear structures of known type. Thus, in order to test the usefulness of the MD and KLD approach to recognise changes occurring in complex processes, we used different model datasets as well as real-world time series.

We started from the testing of sensitivity of MD and KLD measures in case of short datasets generated by several low dimensional attractors like Lorenz and Henon with the added white noises. According to earlier reports adding of noise influences clustering characteristics of compared groups (Vorraboot et al., 2015). Next, we proceeded to the analysis of the original real-world time series. Namely, as mentioned above, we used internet, seismological, meteorological, physiological and economic datasets. It is important to note that, earlier analysis of these datasets indicated the presence of quantifiable nonlinear signatures in their dynamical properties. Thus, we aimed to learn whether these processes (or considered time series) could be regarded as similar or dissimilar by measures provided by MD and KLD metrics.

2.2 Used Datasets and Methods of Analysis

In this research, we used two types of low-dimensional models as well as five types of real-world datasets to demonstrate how, in the case when only short- and low-quality datasets are available, the similarity testing can be used to judge dynamical features of considered processes.

We used 200–500 data long time series of X components of Lorenz and Henon attractors. Generally, Lorenz system (1) is a model describing very complex motion of an incompressible fluid contained in a cell, when the temperature at the bottom is higher and is lower at the top (see e.g., Abarbanel et al., 1993; Hilborn, 1994):

$$\frac{dx}{dt} = p(y - x),$$
$$\frac{dy}{dt} = -xz + rx - y, \quad (2.1)$$
$$\frac{dz}{dt} = xy - bz.$$

Here, p, b and r are system parameters. For present research purposes we assume $r = 0.7$. Used time series have been generated by the discrete version (2) of the Lorenz equations modified by introducing two random noises:

$$x_{t+\Delta t} = p(y_t - x_t)\Delta t + x_t + c\xi_t + \varepsilon\zeta_x,$$
$$y_{t+\Delta t} = (-x_t z_t + rx_t - y_t)\Delta t + y_t + c\xi_t + \varepsilon\zeta_y, \quad (2.2)$$
$$z_{t+\Delta t} = (x_t y_t - bz_t)\Delta t + z_t + c\xi_t + \varepsilon\zeta_z.$$

The role of noise, ξ, is to keep states of the system around the attractor in the origin (0, 0, 0). The role of second noise ζ_x (ζ_y and ζ_z) which is generated for each of the three equations is to increase randomness in the system. For generation of time series, we assume the following values for parameters, $p = 10$, $r = 0.7$, $b = 8/3$, $c = 3$, the initial values $(x(0), y(0), z(0)) = (0, 0, 20)$, and the time step $\Delta t = 0.001$ (see details in Matcharashvili et al., 2019).

Henon attractor or Hénon system (3) is a model of the dynamics of stars moving within galaxies:

$$X_{k+1} = Y_k + 1 - \alpha X_x^2,$$
$$Y_{k+1} = \beta X_k, \quad (2.3)$$

α and β parameters define the type of dynamical research (see e.g. Abarbanel et al., 1993). According to research purposes, white noise of increased intensity has been added to this model system.

Next, we proceed to the real-world datasets. Exactly, we used Border Gateway Protocol (BGP) datasets collected in the frame of the Route Views project (http://www.routeviews.org/). These datasets of four Internet Service Providers, such as AT&T, NTT, IIJ, and Tinet are freely available at: http://figshare.com/articles/Correlation_in_global_routing_dynamics/1549778. For our research purpose from all available databases, we selected just 140-day long part. Next, from this 140-day long time series recorded at mentioned four AS-es, we selected smaller parts which, according to certain analysis, have been recognised as most random-like and less random-like (more regular) (details of BGP data preprocessing and selection can be found in Matcharashvili et al., 2020a and 2020b).

Analysed in this research geophysical datasets have been obtained from southern California earthquake catalogue, at M2.6 representative threshold. Exactly, sequences of times between consecutive earthquakes and distances between them (mentioned above as IET and IED datasets) have been analysed (see details in Matcharashvili et al., 2018 and 2019).

Meteorological datasets represent the yearly number of days when anomalies of daily max temperatures (in Tbilisi, Georgia) significantly deviated from the mean of anomalies for those days (in period from 1914 to 2014). Exactly, as mentioned above, we analysed datasets of yearly number of days, when anomalies of max daily temperatures were significantly larger or were significantly lower than mean values calculated for analysed period (see for details Matcharashvili et al., 2017 and 2018). These original time series were compared with the corresponding randomised meteorological datasets – RAMaxT and RAMinT.

Physiological datasets represented arterial systolic and diastolic blood pressure time series as well as heart rate time series. Exactly, we considered these time series from 70 volunteers of healthy persons and patients with the arterial hypertension. Used in this research physiological datasets have been collected at Institute of Clinical Cardiology, Tbilisi, Georgia. The monitoring of blood pressure was carried out from 12.00 AM to 12.00 AM of the next day. These data have been obtained from 24-hour ambulatory monitoring recordings at 15-minute sampling time (Janiashvili et al., 2013; Matcharashvili et al., 2017).

Economic datasets such as components of Index of Economic Freedom (IEF) [according to Heritage Foundation Report (http://www.heritage.org/index)] and doing business index (DB) [according to World Bank rankings (https://www.doingbusiness.org/en/rankings/)] of three southern Caucasian countries obtained from corresponding international databases.

As we already mentioned, former analysis, carried out on all these datasets, indicated the presence quantifiable nonlinear structure in their dynamical properties. Thus, we aimed to find out whether these processes (or considered time series), are similar or dissimilar by the measures provided by MD and KLD metrics.

In general, the main objective of a good clustering algorithm is to compare classes (or clusters) and to group them using a divergence measure as a metric of similarity or dissimilarity between them (Haykin, 1999; Martins et al., 2004). As said above, for this purpose in present research we used the KLD and the Mahalanobis distance (McLachlan, 1999; Haykin, 1999; Martins et al., 2004). Both these measures incorporate the spatial statistics of the data, giving us a good measure of the distribution of the points, making possible the algorithm to be used to classify very complex datasets. In common parlance, the similarity (or dissimilarity) is assessed by a number that measures how distinct datasets are. The greater (lower) the number of similarity (dissimilarity) measure is, the less (more) distinct the datasets are. It can be added here that the Mahalanobis distance is convenient for our purpose because in contrast to the Euclidean distance the MD is taking into account the correlations of the dataset.

The Mahalanobis distance is defined as:

$$D^2 = (\langle a_i \rangle - \langle b_i \rangle)^T S^{-1} (\langle a_i \rangle - \langle b_i \rangle), \qquad (2.4)$$

where S is the covariance matrix of the distribution of the probability that represents the spatial statistics (Mahalanobis, 1930; McLachlan, 1999). Generally, two states of systems are more probable to fall in the same class or group (or are similar at higher probability) when calculated MD value is smaller. Mahalanobis distance metric can adjust the geometrical distribution of data so that the distance between similar data points is small (Xing et al., 2003; Xiang et al., 2008). Thus, it can enhance the performance of clustering or classification algorithms (Xiang et al., 2008). Besides MD, to measure the statistical 'distance' between the distributions p and q, for a given random variable X, it is common to also use KLD (Cover and Thomas, 2006). It is also known as relative entropy, and is defined as:

$$D_{KL} = \int_{k \in X}^{n} p(x) \log_2 \frac{p(x)}{q(x)}. \qquad (2.5)$$

KLD (D_{KL}) is a measure based on the relative entropy of two probability density functions built for considered processes. Also, the KLD is the symmetric divergence between two classes of compared groups or datasets. In other words, the KLD represents a measure of degree of difficulty of discriminating between classes (the larger the divergence, the greater the separability between the classes).

As it was already mentioned, the use of distance measures in Statistics, including MD and KLD, is of fundamental importance in solving different practical problems, such as hypothesis testing, goodness of fit tests and especially for solving classification tasks.

We point again that, the main objective of this research represented classification of compared groups by certain dynamical features of the considered processes.

After, according to the targeted research purpose, we compared the original data sequences with the surrogate time series in which dynamical structure has been intentionally destroyed (Kantz and Schreiber, 1997; Matcharashvili, 2017). Such comparison helps to understand whether observed changes are indeed connected with the internal dynamical structure of the analysed process.

2.3 Results and Discussions

As it was already mentioned above, in this research, time series of model systems as well as time series of different real-world processes have been analysed.

We have started from the analysis of short time series of X components of Lorenz and Henon attractors with different amount of added noises.

In Figures 2.1 and 2.2, we present results of MD and KLD measures calculation for relatively short (200–500 data) Lorenz and Henon time series. Here, original Lorenz and Henon time series are compared with the same time series in case when increased intensity of noise (see columns 1–8 in Figure 2.1) is added to the original model system. From these figures, it is clear that calculated KLD and MD values increase in parallel to the rise of added noise intensity. This means that increased extent of noise in model time series leads to the increase of difference between compared time series and is in agreement with earlier findings about influence of noise on the separability of compared groups (Vorraboot et al., 2015). In other words, obtained results

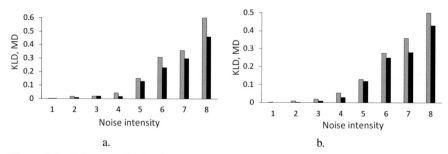

Figure 2.1 Influence of noise intensity on calculated KLD (grey columns) and MD (black columns) values of time series of X components of Lorenz system (a) and Henon attractor (b).

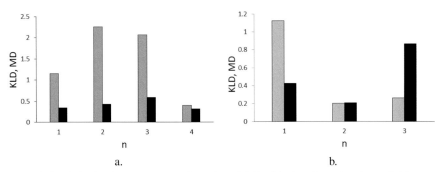

Figure 2.2 Calculated KLD (grey columns) and MD (black columns) values of (a) comparison of regular and irregular parts of BGP updates time series ATT(1), NTT(2), IIJ(3) and Tinet(4); (b) comparison of regular parts of BGP updates time series recorded at ATT AS with BGP updates time series recorded at Tinet(1), NTT(2) and IIJ(3) AS-es.

points that when the noise is added (which logically leads to the decrease of regularity) the model process becomes more and more different from the original one. It is worth mentioning that according to measured classification values, the difference is noticeable even for small noise intensities. This points that used classification measures enable to differentiate small dynamical changes in the model time series even for relatively small datasets (200–500 data). At the same time, we see that calculated KLD values prevail MD values. It is also worth mentioning here that compared two distributions are different when KLD is larger than zero while MD value should fulfill stronger significance requirements in order compared groups to be regarded as different (see e.g. Darscheid et al., 2018; McLachlan, 1999).

After, we proceed to the analysis of the real-world measurement time series. First group of such time series was the BGP dataset. Exactly, as said in Section 2.2, we analysed smaller parts of available long BGP data series which according to previously carried out analysis, have been recognised as most random-like and/or less random-like (more regular) parts in the entire datasets of four different AS-es (see details in Matcharashvili et al., 2020a and 2020b).

In Figure 2.2a, we present results of comparison of regular and irregular parts of BGP updates time series. We see that regular parts of BGP updates time series recorded at ATT, IIJ, NTT and Tinet AS-es are different from irregular parts of corresponding time series. This is clear from calculated values of KLD measures where we observe large measured values indicating that compared clusters (groups) are different. Calculated MD values generally

2.3 Results and Discussions 27

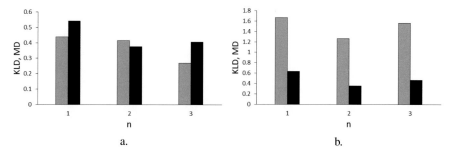

Figure 2.3 KLD (grey columns) and MD (black columns) values of time series of, (a) IET (normed inter earthquakes times) compared with RNT (normed randomised inter earthquakes times), (b) IED (normed inter event distances) compared with RND (normed randomised inter event distances).

confirm this finding though the difference is not always as obvious as for obtained KLD values. This once again indicates found earlier differences in the dynamical features of these two types of BGP updates process (i.e. regular and random-like variability of BGP updates process). At the same time, it needs to be underlined that, the mentioned difference between regular and irregular parts of BGP updates variability is different in different cases. Example, KLD and MD measures values are smaller when regular and irregular parts of Tinet BGP datasets are compared. This points that, in this case, BGP updates process is less different from randomness, opposite to cases observed for other AS-es.

It was also shown that extent of regularity of the process of BGP updates recorded at different AS-es may be different. Indeed, according to Figure 2.2b, regular parts of BGP updates time series recorded at ATT AS, compared with BGP updates time series recorded at other AS-es: IIJ, NTT and Tinet, are different. It is remarkable that regular part of the BGP updates recorded at NTT AS seem is less different from ATT, than regular parts of BGP updates process from IIJ and Tinet AS-es. This can be explained by the fact that, close to be regular BGP updates processes from ATT and NTT As-es are clearly different from randomness while in case of Tinet AS this difference is not so obvious (see Figure 2.2a). Here, we also mention that difference between compared groups in most cases is better visible by calculated KLD values and just in one case MD value prevail (Figure 2.2b, case 3).

Next type of real-world measured time series, which we considered, is from seismology. Namely, as it was explained in Section 2.2, we analysed IET and IED datasets normed to corresponding standard deviations.

In Figure 2.3, we show results of calculation for IET and IED seismological datasets. As follows from this figure, the original dataset of inter earthquake times as well as inter earthquake distances assessed for small segments of original datasets show differences indicated by non-zero KLD values. Calculated MD values also show sign of difference though, again, the difference is less obvious and insignificant. It is important that here we compare IET and IED datasets in the period prior to strongest catalogued earthquakes. According to present views, these periods are characterised by lower extent of regularity compared to the entire seismic cycle including earthquake preparation and post-earthquake aftershock activity (Matcharashvili et al., 2018, 2019). Thus, strong differences between original and randomised seismological datasets generally were not expectable. At the same time, recent findings in the earthquake generation dynamics give basis to suppose that seismic processes prior to strong earthquake occurrence should not be completely random and apparently, it reveal some extent of long range correlations (Lennartz et al., 2008; Corral et al., 2008; Matcharashvili et al., 2019). Presented in Figure 2.3, results are in general agreement with such views, showing that compared to random processes, there are some differences between time and space distributions of seismic processes prior to strong earthquakes. Thus, for geophysical datasets, we also observe efficiency of used classification approaches to differentiate dynamical changes occurred in the seismic process, using just short datasets.

Next, we proceed to the analysis of meteorological datasets. As mentioned in Section 2.2, we consider datasets of yearly number of days with significantly larger – AMaxT, or significantly lower – AMinT, daily temperatures relative to corresponding mean. As it follows from Figure 2.4 (1 and 2), AMaxT and AMinT time series are weakly different from corresponding randomised datasets. Contrary to MD, calculated values of KLD are small but still non-zero and thus we can speak about, at least, weak differences between observation (original) and randomised groups. At the same time, KLD and MD values are essentially larger when we compare AMaxT with AMinT datasets. This means that, found earlier difference between variability features of the yearly number of warmer and colder days (see Matcharashvili et al., 2017) is confirmed by the classification testing based on shorter time series. It is interesting that two datasets hardly different from randomised counterparts, appeared clearly different between each other, as follows from values calculated by used classification methods (compare 1 and 2 cases with case 3 in Figure 2.4).

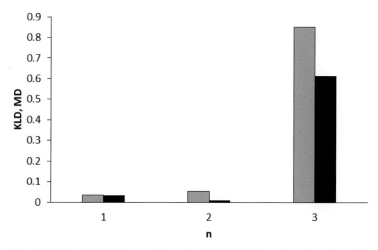

Figure 2.4 Calculated KLD (grey columns) and MD (black columns) values of meteorological time series. (1) Original AMaxT compared with randomised RAMaxT. (2) Original AMinT compared with randomised RAMinT. (3) AMaxT compared with AMinT.

Results of analysis carried out on physiological datasets are presented in Figures 2.5 and 2.6. As it was said in Section 2.2, we analysed arterial systolic and diastolic blood pressure as well as heart rate datasets of persons falling into four arterial hypertension grades, according to guidelines of European Society of Hypertension and the European Society of Cardiology, published in 1997. These physiological characteristics of patients from different hypertensive categories (optimal, normal, high-normal and hypertension) are often discussed in the special scientific literature (see e.g. Pastor-Barriuso, 2003; Janiashvili et al., 2013; Matcharashvili et al., 2018b).

As we see in Figure 2.5, the character of variability of systolic and diastolic blood pressure time series are different in all categories. This is most noticeable from calculated values of KLD measure. Generally, the same can be concluded from obtained MD values when comparing short systolic and diastolic blood pressure time series.

Further, we compared the character of variability of systolic and diastolic blood pressure as well as heart rate time series in the optimal category with the character of variability in other categories (i.e. normal, high-normal and hypertension categories). In Figure 2.6, we see that variability of all considered physiological time series, in normal, high-normal and hypertension

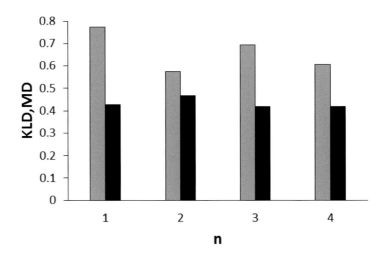

Figure 2.5 Calculated KLD (grey columns) and MD (black columns) values of comparison between systolic and diastolic blood pressure time series: optimal (1), normal (2), high-normal (3) and hypertension (4) categories.

categories is clearly different compared with the optimal category. These differences are most noticeable according to calculated values of KLD measure. As for calculated MD values, changes are not so clear especially for heart rate variability time series presented in Figure 2.6c.

As it follows from the results of our analysis in Figures 2.5 and 2.6, conclusions based on KLD and MD values generally are similar. At the same time, knowing how important is the knowledge of the character of changes in used physiological characteristics it becomes obvious that for such a short (200 data) time series it is preferable to rely on the results of KLD calculation.

The last group, we analysed in this research are economic datasets. Exactly components of Index of Economic Freedom and doing business index datasets, obtained from corresponding international databases. For our research needs, we selected IEF and DB index datasets for south Caucasian Countries: Azerbaijan, Armenia and Georgia in the period 2014–2020.

In general, an index of economic freedom is a measure involving countries' sub-indicators such as trade freedom, tax burden, judicial effectiveness, etc. These indicators when they are weighted according to their influence on economic freedom can be compiled into a single score that is used for a ranking of each Country. IEF is one of the so called composite indicators which by most experts is recognised as essential for different research, social

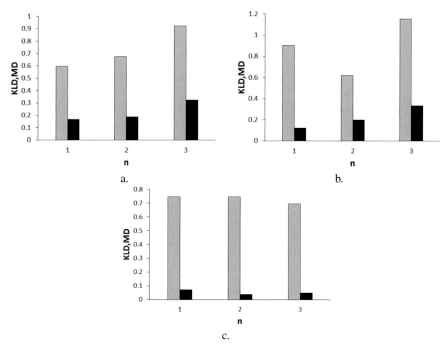

Figure 2.6 Calculated KLD (grey columns) and MD (black columns) values of comparison of optimal blood pressure category with normal (1), high-normal (2) and hypertension (3) categories. Systolic (a), diastolic (b) blood pressure and heart rate (c) time series.

and economic purposes, though still there are literature sources criticising it from methodological matters.

As we see in Figure 2.7, comparing by sub-components of IEF, three south Caucasian countries according to KLD values show that they are different for all considered periods, 2014–2020. Difference according to calculated MD values are not so obvious, though these results at least do not contradict to conclusions based on obtained KLD values. It is also noticeable that observed differences are varied in time. Namely, by variability of sub-components of IEF Armenia vs. Azerbaijan looks less different in 2019 and 2020 (Figure 2.7a). The same can be said about the comparison of Georgia and Azerbaijan in 2019 and 2020 (Figure 2.7c). Interesting is the situation when by the variability of sub-components of IEF index we compare Georgia with Azerbaijan (Figure 2.7b). In this case, in 2019 and 2020, the difference is clear by both calculated KLD and MD values.

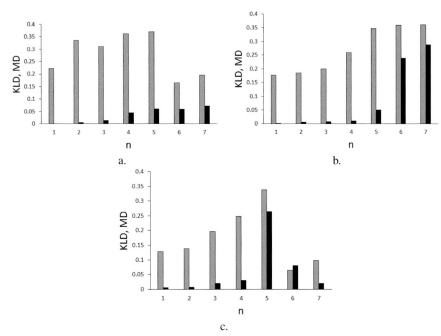

Figure 2.7 Calculated KLD (grey columns) and MD (black columns) values for south Caucasian Countries when they compared by datasets of sub-components of IEF for the period from 2014 to 2020 (1–7 in abscissa axis); (a) Armenia versus Azerbaijan, (b) Georgia versus Armenia, (c) Georgia versus Azerbaijan.

Next, considered economical characteristic – doing business index is also composite indicator. It is an aggregate figure that includes different sub-indicators which define the ease of doing business in a country and is often used for different research and practical purposes.

In Figure 2.8, results of calculations for the datasets of sub-indicators of doing business index are presented. As follows from this figure, till the middle of observation period (2014–2017), Georgia by the variability of these sub-indicators almost similarly is different from Armenia as well as from Azerbaijan. This looks logical in the light of World Bank ranking data (2014–2020) indicating clear difference in DB indexes between Georgia and other south Caucasian Countries. At the same time, since 2017 noticeably changes were observed in the economic development of Azerbaijan and as a result the situation with doing business in Country has been improved according to World Bank ranking database (2014–2020).

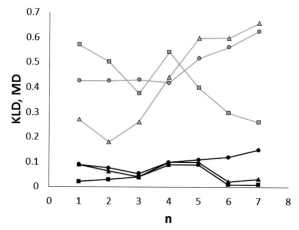

Figure 2.8 Calculated KLD (grey) and MD (black) values for south Caucasian countries when they compared by datasets of sub-components of DB index in period from 2014 to 2020 (1–7 in abscissa axis); Armenia versus Azerbaijan – triangles, Georgia versus Armenia – circles, Georgia versus Azerbaijan – squares.

Thus, is not surprising that the difference between Georgia and Azerbaijan since 2017 began to decrease as it follows from Figure 2.8 (squares). Meanwhile, it follows from the same Figure 2.8, that the difference between Georgia and Armenia (circles) as well as Azerbaijan and Armenia (triangles) continues to increase. All these conclusions are based on the calculated KLD values. Changes by MD values generally do not contradict to what we draw from KLD values, but still are less convincing. Here, it need to be said that used here economic datasets (of several tens of data) are shortest among all considered in this work and thus obtained MD and KLD testing results, without additional analyses, hardly can serve as a basis for strong conclusions. We just showed that when we are not able to get better datasets, the classification assessments, such as used here, may still provide some understanding of the situation which, according to our previous results do not contradict the real state of the case.

2.4 Summary

In this work, we used classification testing based on KLD and MD calculation for the analysis of time series from different model systems as well as datasets of measurements of the real-world processes.

Exactly, we analysed short datasets of Lorenz and Henon attractors with and without added noise. It was shown that gradual increase of noise intensity leads to the increase in the calculated values of MD and KLD. This indicates that even slight changes in the original dynamics, caused by white noise, can be recognised by used separation measure procedures.

Besides of model datasets, real-world measurements of time series, such as Internet, seismological, meteorological, physiological and economic ones, also have been analysed. For all used datasets, it was shown that classification testing helps to recognise slight changes occurred in the variability features in the considered processes. The quality of separation for the used short and mostly imperfect datasets can be named as acceptably good.

Obtained from present analysis, results are in complete accordance with earlier conclusions, carried out on the same or similar longer datasets, based on different linear and nonlinear methods of complex data analysis. According to obtained results, in the case of short-time series, from the point of view of fulfillment of the classification of small dynamical changes, the Kullback–Leibler information-based discrepancies testing in most cases tended to outperform the sensitivity of testing based on the Mahalanobis distance.

Acknowledgments

We would like to thank MSci. N. Zhukova and Dr. I. Davitashvili for the help in the data handling and related technical support.

References

Abarbanel, H.D.I., Brown, R., Sidorowich, J.J., and Tsimring, L.S. (1993). The analysis of observed chaotic data in physical systems. *Reviews of Modern Physics* 65(4), 1331–1392.

Cover, T.M., and Thomas, J.A. (2006). *Elements of Information Theory*, 2nd ed. New Jersey: John Wiley & Sons.

Corral, A. (2008). Scaling and universality in the dynamics of seismic occurrence and beyond. In Carpinteri, A. and Lacidogna, G. (eds). *Acoustic Emission and Critical Phenomena*. London: Taylor & Francis Group, pp. 225–244.

Darscheid, P., Guthke, A., and Ehret, U. (2018). A maximum-entropy method to estimate discrete distributions from samples ensuring nonzero probabilities. *Entropy* 20(8), 601. doi:10.3390/e20080601.

Haykin, S. (1999). *Neural Networks a Comprehensive Foundation*. New Jersey: Prentice Hall.

Hilborn, R.C. (1994). *Chaos and Nonlinear Dynamics: An Introduction for Scientists and Engineers*. New York: Oxford University Press.

Janiashvili, M., Jibladze, N., Matcharashvili, T., and Topchishvili, A. (2013). Comparison of statistical and distributional characteristics of blood pressure and heart rate variation of patients with different blood pressure categories. *Model Assisted Statistics and Applications* 8, 177–184.

Lennartz, S., Livina, V.N., Bunde, A., and Havlin, S. (2008). Long-term memory in earthquakes and the distribution of interoccurrence times. *Europhysics Letters* 81, 69001. doi:10.1209/0295-5075/81/69001.

Mahalanobis, P.C. (1930). On tests and measures of group divergence. *Journal of the Asiatic Society of Bengal* 26, 541–588.

McLachlan, G.J. (1999). Mahalanobis distance. *Resonance* 6, 20–26.

Martins, A.M., Neto, A.D.D., de Melo, J.D., and Costa, J.A.F. (2004). Clustering using neural networks and Kullback-Leibler divergency. In *2004 IEEE International Joint Conference on Neural Networks (IEEE Cat. No.04CH37541), Budapest, Hungary*, pp. 2813–2817, doi:10.1109/IJCNN.2004.1381102.

Matcharashvili, T., Elmokashfi, A., and Prangishvili, A. (2020a). Analysis of the regularity of the Internet Interdomain Routing dynamics. *Physica A*, 124142, 2020. doi.org/10.1016/j.physa.2020.124142.

Matcharashvili, T., and Prangishvili, A. (2020b). Quantifying regularity of the Internet Interdomain Routing based on Border Gateway Protocol (BGP) data bases. In *2020 International Conference on Electrical, Communication, and Computer Engineering (ICECCE), Istanbul, Turkey*. doi: 10.1109/ICECCE49384.2020.9179264.

Matcharashvili, T., Hatano, T., Chelidze, T., and Zhukova, N. (2018a). Simple statistics for complex Earthquake time distributions. *Nonlinear Processes in Geophysics* 25, 497–510.

Matcharashvili, T., Zhukova, N., Chelidze, T., Baratashvili, E., Matcharashvili, T., and Janiashvili, M. (2018b). The analysis of variability of short data sets based on Mahalanobis distance calculation and surrogate time series testing. In: Rojas, I., Pomares, H., and Valenzuela, O. (eds). *Time Series Analysis and Forecasting, Selected Contributions*. Switzerland: Springer Nature Switzerland AG, pp. 275–287.

Matcharashvili, T., Czechowski, Z., and Zhukova, N. (2019). Mahalanobis distance-based recognition of changes in the dynamics of a seismic process. *Nonlinear Processes in Geophysics* 26, 291–305.

Matcharashvili, T., Zhukova, N., Chelidze, T., Founda, D., and Gerasopoulos, E. (2017). Analysis of long-term variation of the annual number of warmer and colder days using Mahalanobis distance metrics – A case study for Athens. *Physica A* 487, 22–23.

Kantz, H., and Schreiber, T. (1997). *Nonlinear Time Series Analysis*. Cambridge: Cambridge University Press.

Pastor-Barriuso, R., Banegas, J.R., Damian, J., Appel, L.J., and Guallar, E. (2003). Systolic blood pressure, diastolic blood pressure, and pulse pressure: An evaluation of their joint effect on mortality. *Annals of Internal Medicine* 139, 731–739.

Vorraboot, P., Rasmequan, S., Chinnasarn, K., and Lursinsap, C. (2015). Improving classification rate constrained to imbalanced data between overlapped and non-overlapped regions by hybrid algorithms. *Neurocomputing* 152, 429–443.

Xiang, S., Nie, F., and Zhang C. (2008). Learning a Mahalanobis distance metric for data clustering and classification. *Pattern Recognition* 41, 3600–3612.

Xing, E.P., Ng, A.Y., Jordan, M.I., and Russell, S. (2003). Distance metric learning, with application to clustering with side-information. In: *Advances in NIPS*. Cambridge, MA: MIT Press, pp. 505–512.

3

A Comparative Study of CNN Architectures for Remaining Useful Life Estimation

Rahul Joshi, Satvik Bhatt, Amitkumar Patil, and Gunjan Soni

Department of Mechanical Engineering, MNIT Jaipur, India
E-mail: {2016ume1466,2016ume1040}@mnit.ac.in

Abstract

In today's technology-oriented industry, maintenance management plays a crucial role by minimising overall production costs and continuously improving system reliability and product quality. Among various maintenance strategies, condition-based maintenance policy has proven to be more effective and efficient with the help of advanced deep learning approaches. A detailed comparison of the established algorithms will provide the management the ability to make prompt and informed decisions, increasing collaboration with site maintenance teams. In previously documented studies, it has been shown that convolution neural network (CNN) offers higher accuracy on smaller datasets. In this study, we aim to evaluate three CNN architectures – LeNet5, AlexNet, VGGNet16 on a set of predefined evaluation metrics, commonly used in the assessment of feature learning algorithms. The C-MAPSS dataset is used as a benchmark to perform our investigation. The paper starts with an overview of data prognostic methods that have seen widespread adoption and gradually converges to explain the sequential techniques used in data pre-processing. An empirical relation between accuracy and time taken to execute is realised. The study shows that the supervised learning model VGGNet16 excels the other models in estimation of engine remaining useful life by yielding a mean accuracy of 95.55%. Transformation of time series representations into recurrence plots enables the network to adapt reliably to the highly varying regression data and produce a good comparative result.

Keywords: condition-based maintenance (CBM), convolution neural network (CNN), prognostic health management (PHM), recurrence plots.

3.1 Introduction

In contemporary engineering services, increasing complexity of systems makes it difficult to detect and perform reliable maintenance activities through conventional methods such as scheduled and breakdown maintenance. The former, grounded on the premise that all equipment follow the bathtub curve model for failure, is found not to be observed in many equipment's (Fu et al., 2004). It is projected that by the year 2026, the global MRO market would account for $40 billion, rising exponentially at a rate of 4.1% yearly (Li et al., 2019). This necessitates the need for a more robust and viable option to undertake maintenance decisions. With the advent of signal processing and ICT, monitoring real-time condition, referred to as condition-based maintenance (CBM), is not only practical, but also feasible.

The remaining useful life (RUL) of a system is predicted based on values obtained from sensors monitoring the operation and is contingent upon the operational settings in which the system is functioning (Kalgren et al., 2006). Several researchers have provided a methodical review of health management in machinery, frequently cited as prognostic health management (PHM). Lei et al. (2018) provided a detailed four-stage layout from data acquisition to estimation of RUL. The trailing work also delineates modelling of health indicators (HI) and classification into appropriate health stage (HS) divisions according to the extent of degradation. A data processing layer is included to account for inherent noise in raw sensor data, before further analysis is carried out in (Jardine et al., 2006; Das et al., 2011). Atamuradov et al. (2017) suggested using a set of factors to perform an assessment of the system components with a high likelihood of failure. However, the procedures involved in acquiring data and extracting desirable features before applying suitable prognostic algorithms remains beyond the scope of this study.

Any CBM practice comprises of two modules – diagnostics and prognostics. Prognostics is responsible for predicting the RUL values, which is associated with cost of maintenance and system reliability (Peng et al., 2010). In general, prognostic techniques are broadly categorised as model-based (Li et al., 2016), data-driven and hybrid (Zhao et al., 2017; Xu et al., 2013) approaches. Model-based approaches, also known as physics of failure (PoF) approaches, create mathematical representations to evaluate physical properties that are responsible for inducing fault in the system. These approaches moreover require a comprehensive knowledge of the system. Further, the harsh conditions in which the fielded systems operate cause unpredictable failure modes that cannot be modelled accurately. Data-driven methods have become increasingly prevalent as a prognostics evaluation tool due to the availability of run-to-failure datasets via open-source data repositories and promising results obtained by machine learning algorithms on prognostics challenges.

Several feature-learning algorithms proposed in literature have attempted to better the accuracy offered by prevailing state of the art techniques. Logistic regression (Lua et al., 2019; Yu, 2017), SVM (Ordóñez et al., 2019; Nieto et al., 2015) and ANN (Laredo et al., 2019; Jain et al., 2014) have successfully been employed in approximating RUL values with significantly low error rates. Neural networks have performed comparatively better due to their ability to represent non-linear functions of the input layer (Szegedy et al., 2014). An et al. (2015) demonstrated the efficacy of neural networks when large labelled datasets are available for complex systems with significant noise. The issue of determining which learning algorithm would produce better performance for a given application has been addressed by the implementation of ensemble learning (Li et al., 2019). Interpretability (Kraus & Feuerriegel, 2019) is a major drawback of conventional shallow network architectures, deterring practitioners from verifying their expertise on management. Bengio (2009) showed how insufficient depth could affect learning of the network.

In the year 2006, a new branch of representation learning, namely deep learning, garnered interest. Difficulty in training deep architectures, specifically neural networks with more than one hidden layer, prevented its usage in real-world applications. Existing optimisation algorithms, for instance, gradient descent, failed to converge to a good solution due to poor initialisation (Bengio et al., 2006). In the seminal work presented by Hinton and Salakhutdinov (2006), an improvement to traditional dimension reduction method, i.e. PCA, was devised, allowing deep networks to be trained without being stuck near local solutions. Two fundamental properties that make deep algorithms attractive are high-level abstraction and reusing of parameters (Bengio et al., 2013). Schmidhuber (2015) provided an overview of deep learned neural networks. Besides the availability of large datasets, surge of interest in the field of deep learning can be attributed to increasing computational power (i.e. GPU unit), advanced research on machine learning, and ease of procuring associated hardware (Deng, 2014).

Among the various deep learning models, convolution neural networks (CNN) is appropriate for handling multi-dimensional data due to convolutional pooling of layers (Zhao et al., 2016). Notably, deep CNN have been used in the prediction of RUL in the following studies (Babu et al., 2016; Zhao et al., 2019; Lia et al., 2018). Additionally, there are only few articles comparing the performance of different learning algorithms. Mathew et al. (2017) compared 10 machine learning algorithms, whereas Li et al. (2019) provided a thorough appraisal of ANN and statistical models for their prediction accuracy. However, to the best of our knowledge, a comparative study on the various CNN architectures (i.e. LeNet-5, AlexNet and VGGnet16) for the estimation of RUL has not been conducted. In this paper, we validate the performance of each architecture based on the following set of parameters – accuracy, loss, recall and precision. This review will help select a suitable model for prognostics, which invariably reduces computation time allowing for a highly efficient and reliable estimator.

3.2 CNN Architectures and Hyperparameters

Many different architectures are being used for predictive analysis. While this paper specifically focusses over three major architectures i.e. LeNet, AlexNet and VGGNet16, all these have their own area of expertise with predictive maintenance being the mutual area of interest.

LeNet-5 (Lecun et al, 1998), represented in Figure 3.2a is composed of five layers out of which two are convolutional and three fully connected layers making it a feed forward neural network. Images with pixel size 50×50 are fed into the input layer in the form of multiple time series windows which is discussed later in this paper. Earlier, Tanh and Sigmoid were used as an activation function, but due to faster training rate (Szandała, 2020), rectified linear unit (ReLU) became more pertinent as an activation function for the hidden layers, followed by SoftMax for output layer. The input layer dimensions are given as: $50 \times 50 \times n$ where n is the number of features. Since, multiple sub-datasets are used, the value of n may vary. In the first two convolutional layers, 32 and 64 filters are used with size 5×5 subsequently max-pooling layers of size 2×2. The last fully connected layer has three nodes denoting three corresponding labels that need to be classified with better results.

AlexNet, first introduced in (Alex Krizhevsky, 2012), won the ImageNet Large Scale Visual Recognition Competition (ILSVRC) 2012. Since then, it has been used extensively as a more capable model for recognising the

3.3 Numerical Experiments   41

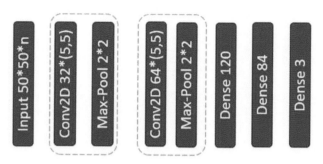

Figure 3.2a   LeNet-5.

patterns in images accurately because of large number of kernels. It consists of eight layers – five convolutional and three fully connected – Figure 3.2b, with 100M parameters. Deployed over the given dataset with a dropout ratio of 0.05 instead of regularisation to eliminate overfitting. Like LeNet, ReLU and Softmax, it is used as the activation functions for hidden layers and output layer. While the first convolution layer uses 32 filters of 11×11, the remaining ones use higher number of filters with size of 5×5 followed by max-pooling layers of 3×3.

Another approach in the domain of deep learning neural network is Visual Geometry Group (VGG). In (2015), Simonyan displayed a network comprising 16 layers of which 13 were convolutional and 3 fully connected. Filters of smaller sizes 3×3 and 2×2, as opposed to AlexNet, with a stride of two were used for convolutional and pooling layers respectively.

The hyperparameters used in the proposed models discussed above are provided in a tabular format in Table 3.1, where SGDM stands for Stochastic Gradient Descent Momentum.

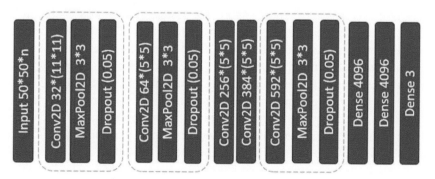

Figure 3.2b   AlexNet.

42 A Comparative Study of CNN Architectures

Figure 3.2c VGGNet16.

Table 3.1 Selected hyperparameters for networks

Optimiser	SGDM/Adam
Learning Rate	0.001
Exponential Decay Rate for 1st Momentum	0.99
Exponential Decay Rate for 2nd Momentum	0.999
Epochs	10
Batch Size	128

3.3 Numerical Experiments

A classification approach is followed to predict the RUL of engine. It requires data in the form of images. Labelling of data is done for identifying the raw data and to set a context so that the proposed models in the literature can learn it. To do so, pre-processing is performed on the regression dataset, which is discussed in the ensuing sections.

3.3.1 Dataset

Simulated data gathered from NASA Prognostics Center of Excellence Data Repository, consists of multi-feature time series data, which makes it a perfect choice while selecting any dataset for comparing the deep learning models. There are 24 features in total, 3 operational settings and 21 sensor measurements shown in (Saxena et al., 2008), having considerable effect on system operation. Every engine function normally without any fault in the beginning which later tend to develop some fatigue and creep and starts to deteriorate gradually with time. The given dataset accounts for initial wear owing to manufacturing inconsistencies (Ramasso & Saxena, 2014). To apply suitable feature learning algorithms and evaluate these data mining models, a fractional value of 0.2 (80%–20% split) is chosen for train-test split to divide the dataset into training and testing. All the engines in the test data are stopped

Table 3.2 Twenty-one sensor recordings with three operational settings

S.No.	Description	Range
Operational Settings		
1	Altitude	0–42K ft.
2	Mach Number	0–0.84
3	Throttle resolver angle	20–100

S.No.	Description	Units
Sensor Measurements		
S1	Inlet fan total temperature	°R
S2	LPC outlet total temperature	°R
S3	HPC outlet total temperature	°R
S4	LPT outlet total temperature	°R
S5	Fan inlet pressure	Psia
S6	Bypass-duct net pressure	Psia
S7	HPC outlet net pressure	Psia
S8	Physical fan speed	Rpm
S9	Physical core speed	Rpm
S10	Pressure ratio of Engine (P50/P2)	-
S11	HPC outlet static pressure	Psia
S12	Fuel flow to Ps30 (pps/psi) ratio	-
S13	Corrected fan speed	Rpm
S14	Corrected core speed	Rpm
S15	Bypass Ratio	-
S16	Fuel-air ratio for burner	-
S17	Bleed Enthalpy	-
S18	Demanded fan speed	Rpm
S19	Demanded corrected fan speed	Rpm
S20	HPT coolant bleed	lbm/s
S21	LPT coolant bleed	lbm/s

at an instant before complete failure to not only have operational cycles different from training datasets, but also getting the base for truth values in the bucket. The objective is to compare the performance of different deep learning models, which further help in predicting the number of remaining functioning cycles more efficiently in the test set.

3.3.2 Pre-processing

Labelling and normalisation – using Equation (3.1) – of regressed data are required for the consistent flow of input and output vectors to the architecture. Finally, time series data are converted into images with the help of recurrence plots prior to deploying CNN networks for classification purposes. Following

Figure 3.3a Steps in pre-processing.

Table 3.3 Labelling of dataset

Cycle Range (RUL)	Label
0–15	2
16–45	1
>45	0

flowchart, Figure 3.3a, provides an overview of the steps involved in pre-processing of raw data.

3.3.2.1 Labelling

For labelling, the entire dataset is divided into three categories (Table 3.3).

It can be speculated that the class labeled as 2 is the most economically valued as it has a smaller number of remaining useful cycles. So, predicting this class with good performance metrics will allow the author to keep apart the best model out of the group. Ultimately, it will serve its purpose of running a smooth and cost-efficient maintenance programme without any havoc and future faults.

3.3.2.2 Scaling of data

Since the data has a huge range for each feature as well as the numeric values are widespread, they are incomparable. Thus, to keep those values in a similar range, we used normalisation. For scaling purpose, we utilised the following equation:

$$Norm(a) = \frac{a - min(a)}{max(a) - min(a)} \qquad (3.1)$$

This yields all values in the range of [0, 1]. As the data gets closer now it is simpler to use for further computation.

3.3.2.3 Splitting of data

With the intention of having a large training dataset, a sliding step of one cycle is used to fragment the time series with a fixed window. Table 3.4 depicts a similar window-wise segregation of 50 cycles corresponding to a given engine.

3.4 Results   45

Table 3.4 Window-wise segregation

Window No.	Cycle range
Window 1	Cycle 0 to Cycle 50
Window 2	Cycle 1 to Cycle 51
Window 3	Cycle 2 to Cycle 52
: : :	: : : :
Window 142	Cycle 141 to Cycle 191

In a similar way, a definite number of windows is framed for the rest engines to have each datapoint at our disposal.

3.3.2.4 Time series to recurrence plots

Since CNN works better with computer vision, it would be better if one use images for input time series datasets to feed into classification model. Thus, time series windows (of length 50 cycles) were transformed into images with the aid of Recurrence Plots. Python libraries such as SciPy are imported to execute code which help in constructing these plots. A plot of dimension 50×50 is generated for each time series as shown in Figure 3.3b,c (a depiction of conversion of Altitude regression data to recurrence plot) excluding the time series with 0 variances, producing a $50 \times 50 \times n$ sequence of image for each set of datapoints (n is the number of the time series with non-zero variance). In the same way, the transformation is done for the rest non-zero variance time series.

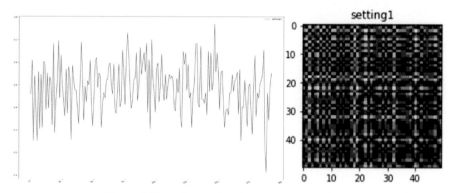

Figure 3.3b,c   Regression data to recurrence plots.

3.4 Results

In this section, a comparative analysis among the mentioned CNN architectures LeNet-5, AlexNet and VGGNet16 is done. These models are trained and tested over two sub-datasets. The results are tabulated as error, accuracy and F1 score as given below. While Error and Accuracy gives idea about the architecture performance, F1 score values tell how good the model is at classifying the labels. Since, the problem is based on 'Imbalanced Classification' a need arises for hyperparameter tuning. Initially these hyperparameters had a fixed value, later trial and error approach is adopted to get maximum accuracy with minimum error.

All the architectures are run on Google Colab (Python 3 Google Compute Engine Backend (TPU) with 35 GB RAM and 177 GB Disk) for 10 epochs while keeping all the hyper parameters tuned to facilitate the possible combinations. Table 3.5 demonstrates comparison results in terms of testing accuracy, error and F1 score as these models are tested on similar datasets. Here, it can be observed that, regardless of the operating conditions, VGGNet16 continually achieves low error values with highest accuracy compared to the other two architectures, while LeNet-5 achieved higher error. The time taken by VGGNet16 was higher relative to LeNet-5 as adding multiple layers to the network also costs time.

Furthermore, a classification report is used to compare results for the projected models with respect to precision, recall and F1 scores. However, only F1 score is used as it is the harmonic mean of recall and precision.

From the tabulated results, it can be observed that the proposed VGGNet16 architecture outscores the existing approaches by producing a relatively higher F1 score, signifying that the failure time predicted via the proposed CNN model is close to their ground truth values. However, if the models are ranked based on time taken to execute, LeNet performs significantly better due to fewer layers. Depending on the application, a trade-off is made between obtaining highly accurate results or reaching a quick solution. When highly accurate results are desired, robust algorithms

Table 3.5 Performance metrices for proposed architectures

Metrics	LeNet-5			AlexNet			VGGNet16		
	Dataset I	Dataset II	Mean	Dataset I	Dataset II	Mean	Dataset I	Dataset II	Mean
Error	0.4	0.28	**0.34**	0.34	0.17	**0.26**	0.31	0.12	**0.22**
Accuracy	92.99	95.14	**94.07**	91.88	93.02	**92.45**	94.65	96.45	**95.55**
F-1 Score	0.68	0.706	**0.69**	0.66	0.706	**0.68**	0.72	0.74	**0.73**
Time (in mins)	0.58	0.86	**0.72**	32.18	43.05	**37.61**	25.18	29.45	**27.31**

– where hyperparameter tuning helps increase network depth – can be implemented to get a closer estimate of the true values. VGGNet provides an excellent option when good results are desired in reasonable time. Moreover, it can be remarked that the performance of these models also got impacted by applying different operating conditions. Analysing each evaluation metric in Table 3.5, it is concluded that better results can be achieved by increasing number of layers and decreasing the size of kernels as an increase in the depth of the network permits it to better learn more intricate features at a substantially lower computational cost.

3.5 Conclusion

In this paper, a comparative study on three well-known CNN architectures is conducted to solve a predictive maintenance problem. In addition to using deep learning for estimation of RUL, the irregularities, and flaws with respect to hyperparameters are also inspected. While estimating RUL of engines, we are wary of the rare events that may occur due to difficulty in collection of such data. Collated data points are used to generate time series in the form of recurrent plots. This helps distinguish engines which are near to the end of their lifecycle. ReLU and Softmax are used as the input and output functions in the above models. The results can be upgraded using proper hyperparameter tuning. The VGGNet16 architecture is truly a pragmatic approach, as it addresses the issue of noise and fault modes occurring in monitoring of various real world PHM applications by providing a reliable value of performance indicators. With evolving deep learning architectures, for our future works, we intend to explore relationships between defined performance metrics and model hyperparameters to help determine RUL values more accurately in the field of PHM. For our future study, we intend to model the degradation of complex systems using state-of-the-art architectures, currently computationally heavy and expensive, to achieve better performance which will result in improved operational design.

References

Alex Krizhevsky, I. S. (2012). ImageNet classification with deep convolutional neural networks. NIPS'12: Proceedings of the 25th International Conference on Neural Information Processing Systems – Volume 1 (pp. 1097–1105). Curran Associates Inc.

An, D., Kim, N. H., and Choi, J.-H. (2015). Practical options for selecting data-driven or physics-based prognostics. Reliability Engineering and

System Safety 133, 223–236. doi:10.1016/j.ress.2014.09.014

Atamuradov, V., Medjaher, K., Dersin, P., Lamoureux, B., and Zerhouni, N. (2017). Prognostics and health management for maintenance practitioners-review, implementation and tools evaluation. International Journal of Prognostics and Health Management 8.

Babu, G. S., Zhao, P., and Li, X.-L. (2016). Deep convolutional neural network based regression approach for estimation of remaining useful life. Database Systems for Advanced Applications 9642, 214–228.

Bengio, Y. (2009). Learning deep architectures for AI. Foundations and Trends® in Machine Learning 2(1), 1–127. doi:10.1561/2200000006

Bengio, Y., Courville, A., and Vincent, P. (2013). Representation learning: A review and new perspectives. IEEE Transactions on Pattern Analysis and Machine Intelligence 35(8), 1798–1828. doi:10.1109/TPAMI.2013.50

Bengio, Y., Lamblin, P., Popovici, D., Larochelle, H., and Montr'eal, U. d. (2006). Greedy layer-wise training of deep networks. Advances in Neural Information Processing Systems 19, Proceedings of the Twentieth Annual Conference on Neural Information Processing Systems. Vancouver.

Das, S., Hall, R., Herzog, S., Harrison, G., Bodkin, M., and Martin, L. (2011). Essential steps in prognostic health management. 2011 IEEE Conference on Prognostics and Health Management. Montreal. doi:10.1109/ICPHM.2011.6024332

Deng, L. (2014). A tutorial survey of architectures, algorithms, and applications for deep learning. APSIPA Transactions on Signal and Information Processing 3. doi:10.1017/atsip.2013.9

Fu, C., Ye, L., Liu, Y., Yu, R., Iung, B., Cheng, Y., and Zeng, Y. (2004). Predictive maintenance in intelligent-control-maintenance-management system for hydroelectric generating unit. IEEE Transactions on Energy Conversion 19(1), 179–186. doi:10.1109/TEC.2003.816600

Hinton, G. E., and Salakhutdinov, R. R. (2006). Reducing the dimensionality of data with neural networks. Science 313(5786), 504–507. doi:10.1126/science.1127647

Jain, A. K., Kundu, P., and Lad, B. K. (2014). Prediction of remaining useful life of an aircraft engine under unknown initial wear. 5th International & 26th All India Manufacturing Technology, Design and Research Conference. Guwahati.

Simonyan, A. Z. (2015). Very deep convolutional networks for large-scale image recognition. International Conference on Learning Representations. arXiv.org.

K. S. Jardine, A., Lin, D., and Banjevic, D. (2006). A review on machinery diagnostics and prognostics implementing condition-based maintenance. Mechanical Systems and Signal Processing 20(7), 1483–1510. doi:10.1016/j.ymssp.2005.09.012

Kalgren, P. W., Byington, C. S., Roemer, M. J., and Watson, M. J. (2006). Defining PHM, a lexical evolution of maintenance. 2006 IEEE Autotestcon. Anaheim. doi:10.1109/AUTEST.2006.283685

Kraus, M., and Feuerriegel, S. (2019). Forecasting remaining useful life: Interpretable deep learning approach via variational Bayesian inferences. Decision Support Systems, 125. doi:10.1016/j.dss.2019.113100

Laredo, D., Chen, Z., Schütze, O., and Sun, J.-Q. (2019). A neural network-evolutionary computational framework for remaining useful life estimation of mechanical systems. Neural Networks, 116, 178–187. doi:10.1016/j.neunet.2019.04.016

Lecun, Y., Bottou, L., Bengio, Y., and Haffner, P. (1998). Gradient-based learning applied to document recognition. Proceedings of the IEEE 86, 2278–2324. doi:10.1109/5.726791

Lei, Y., Li, N., and Liang Guo, N. L. (2018). Machinery health prognostics: A systematic review from data. Mechanical Systems and Signal Processing 104, 799–834. doi:10.1016/j.ymssp.2017.11.016

Li, N., Gontarz, S., Lin, J., Radkowski, S., and Dybala, J. (2016). A model-based method for remaining useful life prediction of machinery. IEEE Transactions on Reliability 65(3), 1314–1326.

Li, R., and Curran, W. J. (2019). Comparison of data-driven prognostics models: A process perspective. 29th European Safety and Reliability Conference. Hannover.

Li, Z., Goebel, K., and Wu, D. (2019). Degradation modeling and remaining useful life prediction of aircraft engines using ensemble learning. Journal of Engineering for Gas Turbines and Power 141(4). doi:10.1115/1.4041674

Li, Z., Goebel, K., and Wu, D. (2019). Degradation modeling and remaining useful life prediction of aircraft engines using ensemble learning. Journal of Engineering for Gas Turbines and Power 141(4). doi:10.1115/1.4041674

Lia, X., Ding, Q., and Sun, J.-Q. (2018). Remaining useful life estimation in prognostics using deep convolution neural networks. Reliability Engineering & System Safety 172, 1–11. doi:10.1016/j.ress.2017.11.021

Lua, F., Wu, J., Huang, J., and Qiu, X. (2019). Aircraft engine degradation prognostics based on logistic regression and novel OS-ELM algorithm. Aerospace Science and Technology 84, 661–671. doi:10.1016/j.ast.2018.09.044

Mathew, V., Toby, T., Singh, V., Rao, B. M., and Kumar, M. G. (2017). Prediction of Remaining Useful Lifetime (RUL) of turbofan engine using machine learning. 2017 IEEE International Conference on Circuits and Systems (ICCS). Thiruvananthapuram. doi:10.1109/ICCS1.2017.8326010

Nieto, P., E.García-Gonzalo, Lasheras, F., and Juez, F. C. (2015). Hybrid PSO–SVM-based method for forecasting of the remaining useful life for aircraft engines and evaluation of its reliability. Reliability Engineering & System Safety 138, 219–231. doi:10.1016/j.ress.2015.02.001

Ordóñez, C., Lasheras, F. S., Roca-Pardiñas, J., and Juez, F. J. (2019). A hybrid ARIMA–SVM model for the study of the remaining useful life of aircraft engines. Journal of Computational and Applied Mathematics 346, 184–191. doi:10.1016/j.cam.2018.07.008

Peng, Y., Dong, M., and Zuo, M. J. (2010). Current status of machine prognostics in condition-based maintenance: A review. The International Journal of Advanced Manufacturing Technology 50, 297–313. doi:10.1007/s00170-009-2482-0

Ramasso, E., and Saxena, A. (2014). Review and Analysis of Algorithmic Approaches Developed for Prognostics on CMAPSS Dataset. Annual Conference of the Prognostics and Health Management Society 2014. Fort Worth.

Saxena, A., Goebel, K., Simon, D., and Eklund, N. (2008). Damage propagation modeling for aircraft engine run-to-failure simulation. 2008 International Conference on Prognostics and Health Management (pp. 1–9). Denver: IEEE. doi:10.1109/PHM.2008.4711414

Saxena, A., Goebel, K., Simon, D., and Eklund, N. (2008). Damage propagation modeling for aircraft engine run-to-failure simulation. Denver: IEEE. doi:10.1109/PHM.2008.4711414

Schmidhuber, J. (2015). Deep learning in neural networks: An overview. Neural Networks 61, 85–117. doi:10.1016/j.neunet.2014.09.003

Szandała, T. (2020). Review and comparison of commonly used activation functions for deep neural networks. Bio-inspired Neurocomputing. Studies in Computational Intelligence 903, 203–224. doi: 10.1007/978-981-15-5495-7_11

Szegedy, C., Zaremba, W., Sutskever, I., Bruna, J., Erhan, D., Goodfellow, I., and Fergus, R. (2014). Intriguing properties of neural networks. International Conference on Learning Representations. Banff.Xu, J., Wang, Y., and Xu, L. (2013). PHM-oriented integrated fusion prognostics for aircraft engines based on sensor data. IEEE Sensors Journal 14(4), 1124–1132. doi:10.1109/JSEN.2013.2293517

Xu, J., Wang, Y., and Xu, L. (2013). PHM-oriented integrated fusion prognostics for aircraft engines based on sensor data. IEEE Sensors Journal 14(4), 1124–1132. doi:10.1109/JSEN.2013.2293517

Yu, J. (2017). Aircraft engine health prognostics based on logistic regression with penalization. Aerospace Science and Technology 68, 345–361. doi:10.1016/j.ast.2017.05.030

Zhao, G., Zhang, G., Ge, Q., and Liu, X. (2016). Research advances in fault diagnosis and prognostic based on deep learning. 2016 Prognostics and System Health Management Conference (PHM-Chengdu). Chengdu. doi: 10.1109/PHM.2016.7819786

Zhao, R., Yan, R., Chen, Z., Mao, K., Wang, P., and Gao, R. X. (2019). Deep learning and its applications to machine health monitoring. Mechanical Systems and Signal Processing 115, 213–237. doi:10.1016/j.ymssp.2018.05.050

Zhao, Z., Liang, B., Wang, X., and Lu, W. (2017). Remaining useful life prediction of aircraft engine based on degradation pattern learning. Reliability Engineering & System Safety 164, 74–83. doi:10.1016/j.ress.2017.02.007

4

The Analysis of Dynamical Changes and Local Seismic Activity of the Enguri Arch Dam

Aleksandre Sborshchikovi, Tamaz Chelidze, Ekaterine Mepharidze,
Dimitri Tepnadze, Natalia Zhukova, Teimuraz Matcharashvili,
and Levan Laliashvili

Ivane Javakhishvili Tbilisi State University, M. Nodia Institute
of Geophysics, Tbilisi, Georgia
E-mail: a.sborshchikov@gmail.com

Abstract

The aim of our research was the analysis of dynamic changes of dam foundation displacement according to the periodic variation water level in the lake around the Enguri Arch Dam. Our database includes the information collected from 1974 to 2020. Dynamical changes of local seismic process from the beginning of the Enguri Arch Dam construction 1974–2020 were tested and analysed by modern several nonlinear methods. For clear data analysis, we choose several methods such as Multifractal Detrended Fluctuation Analysis (MFDFA) and (Lempel-Ziv) Complexity Measure (LZC) as for original also surrogate datasets. From the results of our research, we conclude that dynamic changes of dam foundation displacement are connected with the process of dam behaviour and especially affected by water level change in the reservoir behind the Enguri high dam. According to our results, we can see that the many influences assessment on quantifiable changes in the dynamics of local seismicity around the Enguri Arch Dam.

Keywords: Dynamical changes, nonlinear analyses, datasets.

4.1 Introduction

While studying the literature close to our task, we have found a lot of mentioned facts that the construction and functioning of large water reservoirs may have a strong influence on the environment. Here, we can list some of them: influence on the local seismic activity, changes in local weather, initiation of landslides, etc. Paying attention to the fact that a huge number of researches have been devoted to this question, we can confirm that many aspects of the problem still must be studied in further multidisciplinary research efforts.

This location of the Enguri Dam area was chosen because here engineering research works on the influence of the construction of a high dam and the change of water level in the reservoir on the building itself and the environment were carried out for a long time. During the last century in west Georgia, the construction of a large 271-m high Enguri arch dam has started which still remains as one of the highest (in its class) dams in the world. Nowadays, the Enguri arch dam is the part of the Enguri Hydro Power Plant (HPP) located in the river Enguri Gorge, Georgia. Since the start of construction, the contemporary, to that time, multi-disciplinary geodynamical–geophysical monitoring was organised in the dam area (Chelidze et al., 2013; Chelidze et al., 2020; Chelidze et al., 2021). The geological survey documented that the branch fault of the large active beneath the Enguri dam, Ingirishi fault crosses the right-wing of the dam foundation. At the same time, it is known that the presence of the active (or potentially active) fault in the large dam foundation is a serious threat to dam safety (Chelidze et al., 2013; Matcharashvili et al., 2018; Matcharashvili et al., 2019). It was quite logical that the monitoring of the fault zone started well before the beginning of dam construction and reservoir filling.

The branch fault of the main Ingirishi fault (Figure 4.1) crosses the foundation of the Enguri dam and, thus, poses a significant hazard to the dam.

In fact, the area under study is a natural large-scale laboratory for the investigation of tectonic, man-made and environmental impacts on the fault zone deformation pattern. The summary contributions of these processes are reflected in the time series of fault zone strain. It is evident that the fault dynamic reflects the joint influence of two main factors: one (tectonic strain) leads to piecewise linear (in time) displacement, which we define as a trend component, and the other one to quasiperiodic oscillations, decorating the main trend.

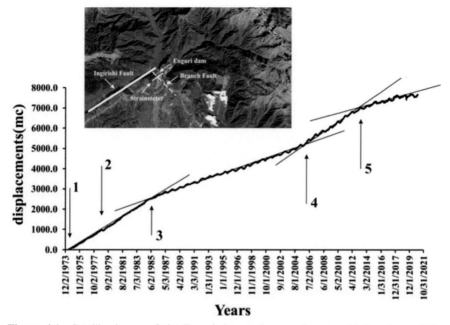

Figure 4.1 Satellite image of the Enguri dam and reservoir area with locations of the main Ingirishi fault and the branch fault, crossing the dam foundation and Dam foundation displacement datasets around EDITA polygon in period from 1974 to 2020. Dam foundation displacement datasets around EDITA polygon in period from 1974 to 2020. Arrows 1–5 correspond to the start of five periods of fault zone extension.

According to the information of this long-period monitoring, nowadays, we have unique databases of some important characteristics, such as high weir foundation and weir body tilts, deformation of foundation, weir body temperature, water level variation in the reservoir, etc. We used these datasets for our researches already published in scientific periodicals (Chelidze et al., 2020; Chelidze et al., 2021; Matcharashvili et al., 2018; Matcharashvili et al., 2019; Peinke et al., 2006). In our study, we aimed to be focussed on dam foundation displacement datasets as well as a seismic catalogue of the area around Enguri arch dam during the period from 1974 to 2020 (Figure 4.1).

Before starting the monitoring, as a precaution measure, the branch fault under the foundation was cleaned from gouge and filled with concrete. Besides, special additional elements (seams and reinforcing) were constructed in the dam body over the fault, which allows withstanding differential displacements up to 10 cm without significant changes in the stress–strain state of the dam. Further, in order to control the fault behaviour permanently,

4 years before the first filling of the reservoir, in 1974, the quartz strainmeter, crossing the fault zone was installed. The strainmeter recorded displacement of the blocks, divided by the fault zone (the thickness of the zone is around 10 m) in the normal to the fault plane direction, so it shows fault zone extension/contraction. The length of the quartz tube is 22.5 m and the free end of the tube is equipped with a photo-optical recording system (now in parallel with a laser system). The tube was placed so that both fixed and free ends are on the opposite sides of the fault zone, separated from it by several meters. The readouts from the photo recording were made once per day at the same time. The displacements sensitivity of the system was of the order of 0.18 m/mm.

Preliminary flooding of the future reservoir territory started at the end of 1977. Since 15 April 1978, the reservoir was filled step by step. Since 1987, the water level in the reservoir has been changing seasonally, almost periodically.

Seismic datasets were obtained from monitoring systems created around the Enguri area. The representative threshold for the local earthquake catalogue from 1974 to 2020 was M2.2.

4.2 Main Text

In our analysis, we take datasets from strainmeter, which fixed measures of displacements of high dam foundation. As we have noted above, filling the Enguri reservoir influences on the seismic activity around of dam. Research was done on interevent sequences (intervals between the events of the earthquakes in minutes) original and surrogate (randomise) seismic catalogue of the Enguri. For investigating the dynamical changes of the dam foundation displacements and seismic behaviour, we have taken several data analysis methods.

For example, multifractal detrended fluctuation analysis (MF-DFA) and algorithmic (Lempel–Ziv) complexity measure.

On the base of Ivane Javakhishvili Tbilisi State University, M. Nodia Institute of Geophysics the software for calculating MF-DFA and LZC was created.

4.2.1 Methods and Results

The first method of our analysis is multifractal detrended fluctuation analysis. MF-DFA method can examine higher-dimensional fractal and multifractal characteristics hidden in time series. However, the removal of local trends in MF-DFA is based on discontinuous polynomial fitting, resulting in pseudo-fluctuation errors.

The MF-DFA was proposed by (Kantelhardt et al., 1951) and is based on DFA (detrended fluctuation analysis). This method is considered to be an effective tool for measuring whether multifractal characteristics exist in seismic and displacement time series (Kantelhardt et al., 1951; Espen, 2012).

We have done this analysis for a different polynomial degree q (so-called fluctuation degree) and determined the multifractal scaling behaviour of time series.

In our study, we have used three main characteristics: Hurst exponent ($H(q)$), Multifractal Dimensions ($D(q)$) and Fluctuation functions ($F(q)$).

The slope H depends on q – order of the polynomial, of the regression line, is called the Hurst exponent (Hurst, 1951). The MF-DFA method determines positive generalised Hurst exponents $h(q)$, and it becomes inaccurate for strongly anti-correlated signals – where $h(q)$ is close to zero. The Hurst exponent defines the monofractal structure of the time series by how fast local fluctuations, grow with increasing segment sample size (scale). The larger Hurst exponent H is visually seen as more slow evolving variations (i.e. more persistent structure) in monofractal and multifractal time series compared with white noise. The Hurst exponent will be in the interval between 0 and 1 for noise-like time series, whereas above 1 for a random walk-like time series.

So, we have a long-range correlated structure – the Hurst exponent is in the interval 0.5–1, an anti-correlated structure – the Hurst exponent is in the interval 0–0.5 and uncorrelated Hurst exponent $H = 0.5$. One more situation is when we have $H = 0.5$, in this case, we can say that we have a short-range dependent structure (Gao et al., 2006).

MF-DFA is directly related to the classical multifractal scaling exponents that generalised multifractal dimension $D(q)$. $D(q)$ depends on the q scale.

Determine the scaling behaviour of the fluctuation functions $Fq(s)$ also depends on values of q and s – time-series scale. Long-range power-law correlated, $Fq(s)$ increases, for large values of s, as a power-law.

The MF-DFA method represents the best method for carrying out multifractal nonstationary time series.

MF-DFA analysis of displacements of high dam foundations shows the behaviour of the complex dynamical structures. The datasets for the period 1974–2020 were fixed for different external influences: filling the reservoir, weather conditions, and seismic activity. Before starting our analysis, we have detrended the displacement datasets. The analysis was done for the polynomial degree p2–p5 (Figure 4.2a–c).

From the results of MF-DFA analysis, we can see how the three main characteristics, $H(q)$, $D(q)$ and $F(q)$ are changed. As we increase the polynomial degree the value of $F(q)$ is decreased. The $H(q)$ is in the numerical range 0.5–1 and above, which indicates that we have a long-range dependent (i.e. correlated) structure of time series. Classical multifractal scaling exponents that generalised multifractal dimension $D(q)$ shows the obstinacy on the dynamical structure of the system.

The next step was MF-DFA analysis of the seismic catalogue around Enguri Arch Dam. For our analysis, we have chosen the optimal period of datasets from 1974 to 2017. For the original one, we have calculated the intervent sequences, so-called waiting times – intervals between earthquakes. After that, we randomised these new datasets 50 times and then we calculate the average and compare the original datasets with the average random database.

We compared the results for polynomial degrees p2 and p5 (Figure 4.3a–f).

In this figure, we can see the results of MF-DFA analysis compare to intervent sequences of the original Enguri seismic catalogue and randomise datasets. The results of the study show us the changes in dynamical structure

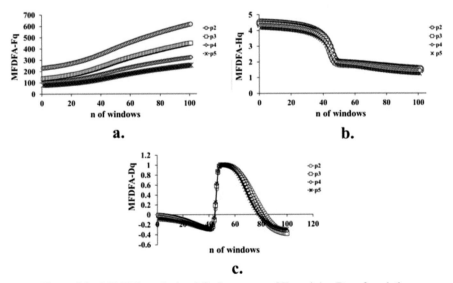

Figure 4.2 MF-DFA analysis of displacements of Enguri Arc Dam foundation.

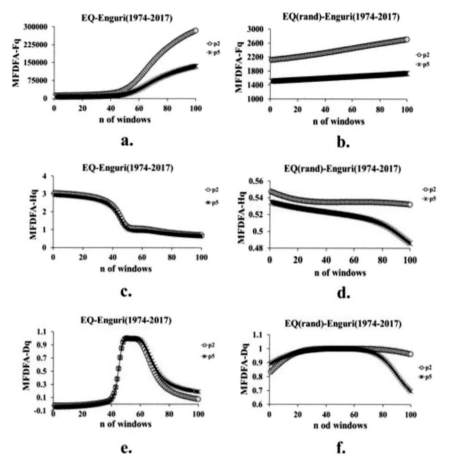

Figure 4.3 MF-DFA analysis of seismic catalogue around Enguri Arc Dam. compare of original interevent sequences with random data.

during the randomisation. From the results of MF-DFA analysis, we can see that $F(q)$ decrease when the order of $p = 5$ for both original and random datasets. The main characteristic $H(q)$ is 0.5–1 and above, which shows long-range dependent (i.e. correlated) structure of time series for original interevent data of Enguri. For randomising interevent data of Enguri $H(q)$ changed in 0–0.5 and 0.5–1. When $p = 2$, we can see $H(q)$ in 0.5–1 – system correlated and when we increase polynomial $p = 5$ $H(q) = 0$–0.5 the dynamical structure is violated and the system becomes anti-correlated. Classical multifractal scaling exponents that generalised multifractal dimension $D(q)$

shows the obstinacy in dynamical structure for original interevent data of Enguri and is destroyed for the randomise interevent data of Enguri.

According to our studies, we can conclude that increasing the degree of polynomial makes changes in the dynamical structure of the system and decreases the fluctuation.

One of the themes of our analysis was using the Lempel and Ziv algorithmic complexity (LZC) calculation (Gao and Zheng, 1993; Gao and Zheng, 1994; Lempel and Ziv, 1976; Hu and Gao, 2006. This method is used for the quantification of the extent of order (or randomness) in datasets of different origins.

The main fact, we should know about LZC is that it transforms the analysed data into a new symbolic sequence. The original datasets are converted into a 0 and 1, sequence by comparing to a certain threshold value (usually median of the original dataset). Once the symbolic sequence is obtained, it is parsed to obtain distinct words, and the words are encoded. Denoting the length of the encoded sequence for those words, the LZC can be defined as

$$C_{LZ} = \frac{L(n)}{n} \qquad (4.1)$$

where $L(n)$ is the length of the encoded sequence and n is the total length of the sequence. Parsing methods can be different (Hu and Gao, 2006; Cover and Thomas, 1991). In this work, we used the scheme described (Hu and Gao, 2006).

LZC was carried out for seismic datasets. For analysis, we used Enguri seismic catalogue for the 1974–2017 period. We have compared the randomised intervent dataset of Enguri seismic data with the original intervent time. For both datasets, we have calculated LZC for windows length 100 max window 3897, and step 100 (Figure 4.4).

According to Figure 4.4 from the results of LZC analysis, we see how the dynamical structure is ordered, when we randomise the interevent data of seismic catalogue.

It is determined that we have the influence on displacement value caused by dam foundation and local seismic activity filing Enguri reservoir, changes in local weather, initiation of landslides, etc.

From the first results of MF-DFA and LZC analysis, we can see how external influences change the dynamical structure of the system and define its behaviour. We hope that our results will help us for future researches and detailed studying of the system behaviour.

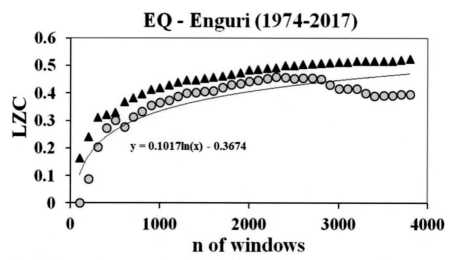

Figure 4.4 Lempel and Ziv algorithmic complexity (LZC) analysis of compare original (grey circles) and randomise (black triangle) seismic data around Enguri Arch Dam.

4.3 Conclusion

From the task of our analysis, we have investigated dynamical characteristics of Enguri high weir foundation displacements together with dynamical changes occurred in the seismic process around the Enguri area. All databases of dam foundation displacements and seismic datasets were collected in period started from 1974 to 2020. In our analysis, we used modern methods of data analysis, such as MF-DFA and LZC. With our research, we conclude that dynamics of dam foundation displacement is strongly influenced by the process of Arc Dam construction and especially by water level change in artificial helps us savour behind Enguri high dam. From MF-DFA and LZC analysis, we got clear results, which helps us to define the degree of dynamical changes and behaviour of our system. It was also shown that the similar factors leaded to quantifiable changes in the dynamics of local seismicity around Enguri Arc Dam.

Acknowlegdements

We want to thank Dr. Irma Davitashvili for the program software.

References

Chelidze, T., Matcharashvili, T., Abashidze, V., Kalabegashvili, M., and Zhukova, N. (2013). Real time monitoring for analysis of dam stability: Potential of nonlinear elasticity and nonlinear dynamics approaches. *Frontiers in Structural and Civil Engineering* 7, 188–205.

Chelidze, T., Tepnadze, D., Mepharidze, E., Sborshchikovi, A., Laliashvili, L., and Matcharashvili, T. (2020). Scaling features of earthquakes occurrences in the equally distributed non-overlapping time windows. *Bulletin of the Georgian National Academy of Sciences* 14(4), 40–45.

Chelidze, T., Matcharashvili, T., Abashidze, V., Dovgal, N., Mepharidze, E., and Chelidze, L. (2021). Time series analysis of fault strain accumulation around large dam: the case of enguri dam, greater caucasus, building knowledge for geohazard assessment and management in the caucasus and other orogenic regions. In: Bonali, F. L., Mariotto, F. P, Tsereteli, N. (eds). *Building Knowledge for Geohazard Assessment and Management in the Caucasus and other Orogenic Regions*. Dordrecht: Springer, pp. 185–204.

Cover, T.M. and Thomas, J.A. (1991). *Elements of Information Theory*. New York: Wiley.

Espen, I.A.F. (2012). Introduction to multifractal detrended fluctuation analysis in Matlab. *Frontiers in Physiology* 3(141), 1–18.

Gao, J.B. and Zheng, Z.M. (1993). Local exponential divergence plot and optimal embedding of a chaotic time series. *Physics Letters A* 181, 153–158.

Gao, J.B. and Zheng, Z.M. (1994). Direct dynamical test for deterministic chaos and optimal embedding of a chaotic time series. *Physical Review E* 49, 3807–3814.

Gao, K., Hu, J., Tung, W.-W., Cao, Y., Sarshar, N., and Roychowdhury, V.P. (2006). Assessment of long-range correlation in time series: how to avoid pitfalls. *Physical Reviews E, Statistical Nonlinear Soft Matter Physics* 73, 016117.

Hu, J. and Gao, J. (2006). Analysis of biomedical signals by the Lempel-Ziv complexity: the effect of finite data size. *IEEE Transactions on Biomedical Engineering* 53, 2606–2609.

Hurst, H.E. (1951). Long-term storage capacity of reservoirs. *Transactions of the American Society of Civil Engineers* 116, 770–799.

Kantelhardt, J.W., Zschiegner, S.A., Koscielny-Bunde, E., Havlin, S., Bunde, A., and Stanley, H.E. (2002). Multifractaldetrended fluctuation analysis of nonstationary time series. *Physica A: Statistical Mechanics and its Applications* 316, 87–114.

Lempel, A. and Ziv, J. (1976). On the complexity of finite sequences. *IEEE Transactions on Information Theory* 22(1), 75–81.

Matcharashvili, T., Hatano, T., Chelidze, T., and Zhukova, N. (2018). Simple statistics for complex earthquake time distributions. *Nonlinear Processes in Geophysics* 25, 497–510.

Matcharashvili, T., Chelidze, T., Zhukova, N., Mepharidze, E., Sborshchikovi, A., Tephnadze, D., and Laliashvili, L. (2019). *Dynamical Changes of Foundation Displacements and Local Seismic Activity Occurred During Construction of Enguri Arch Dam. 25th European Meeting of Environmental and Engineering Geophysics* (vol. 1, pp. 1–5.) The Hague, Netherlands: European Association of Geoscientists & Engineers.

Peinke, J., Matcharashvili, T., Chelidze, T., Gogiashvili, J., Nawroth, A., Lursmanashvili, O., and Javakhishvili, Z. (2006). Influence of periodic variations in water level on regional seismic activity around a large reservoir: field data and laboratory model. *Physics of the Earth and Planetary Interiors* 156, 130–142.

5

Analysis and Prediction of Daily Closing Price of Commodity Index Using Auto Regressive Integrated Moving Averages

Bijesh Dhyani, Manish Kumar, Poonam Verma, and Anurag Barthwal

Management Studies, Graphic Era Deemed To Be University,
Management Studies, Graphic Era University,
Graphic Era Hill University,
Shiv Nadar University,
DIT University, Dehradun, India
E-mail: bijeshdhyani@gmail.com; manishsingh12@rediffmail.com; Poonaddn@gmail.com; ab414@snu.edu.in

Abstract

In most of the developed nations, there are markets where shares, securities or commodities are traded on daily basis, which generally reflects the growth of any country and the health of the company stocks which are traded. Stock market investment is considered to be one of the riskiest investments by investors all around the world, but if historical data is studied carefully then the gap between how the market behaves and what investors know, can be minimized. The data of opening and closing price, which is generated, is a times-series in nature. This time-series data of any index or stock attracts researchers to predict the next move or price of the commodity or index. There are lots of methods available to analyze the data like auto-regressive integrated moving average (ARIMA), regression-based, neural network or moving averages methods. However, ARIMA is a type of model which is generally applied to the time-series data to gain insights into data, to understand what happened in the past and what is the next move to expect. In this chapter, an attempt has been made to use the time-series data of the past 5 years and based on that we can forecast the future direction of the indices.

Keywords: ARIMA, commodity index, ACF, PACF, index analysis, stationarity.

5.1 Introduction

Predicting stock price direction is something individuals and financial firms have researched for years. Books and papers have been written on the subject, but rarely are the results repeatable. Recent research has shown greater predictability in high-frequency stock data (by second or by minute, rather than daily or weekly, but this research is often under-represented in the academic literature. Historically, this is due to the lack of availability of trade-by-trade data and the difficulty in working with such large quantities of it. Furthermore, determining future market direction in practice requires special consideration since streaming stock data may arrive faster than a model may generate results; a model that takes 30 minutes to arrive at a prediction is of little value if the objective was to predict one minute in the future (Aburto et al., 2007). This work explores the predictability of stock market direction using high-frequency stock data.

Stock markets have had a significant impact on many areas like business, education, employment, technology and mainly on the economy. Expert analysts and investors have to develop and testing models on the stock price prediction. Stock prediction is complicated and extremely challenging due to markets dynamics, intrinsic noisy environments, nonlinear, nonparametric, chaotic nature, non-stationary and large volatility with respect to the stock trends. Prediction is complex because stock markets get affected by quarterly earnings reports, economic, political, psychological and company-specific variables, market news and varying changing behaviors (Babu et al., 2012). The two approaches to analyse the stock market prediction are technical and fundamental analysis used to make decisions about stock market prediction. There are several technical indicators based on daily collection trends. Although we collect these on daily basis, it is quite complex to forecast daily and weekly trends in the market. Stock prediction is considered to be a major challenge for increasing production.

It is a known fact that if you are trading in the stock market, the high return you expect is always associated with risk and uncertainty. The awareness of the financial market and the growth of investors has attracted researchers and analysts to forecast the trend and help people in building a portfolio using different models and certainly downsize the risk. In India, there are different indexes like the National Stock Exchange (NSE) and Bombay Stock

Exchange (BSE) where companies are traded and the health of financial markets is monitored. The financial reports which are prepared by utilizing different available models are analyzed and used by investors and companies to enter or exit the market (Mehdi Askari et al., 2011). In India, not only equities are traded, but other things like commodities, company bonds and currency, etc. are also traded (Pai et al., 2005). The use of the ARIMA model is most suited when there is clarity that the time-series data under observation is non-stationary (Anurag Barthwal et al., 2021). Here, non-stationary implies that the time-series has constant mean and variance. In this paper, we have implemented the ARIMA model for time-series data of stocks and prediction is done on the basis of this model. There are many possible approaches to stock market prediction, and most researchers usually stick to one of the following three (Jiang H et al., 2021):

1. *Technical approach*: Within this approach, the stock prices are predicted based on historical information about the price changes. A set of technical indicators calculated upon this data is usually used to characterize the current state of affairs and make a prediction about the future dynamics. One of the most common techniques is finding repetitive trends in the dynamics of the technical indicators and/or historical prices (e.g. 'head and shoulders' pattern).
2. *Fundamental approach*: This approach relies on the assumption that the stock price is independent of past changes and is mostly affected by the events in the outer world as well as the overall economical or political situation. The information that is usually taken into account includes the inflation and unemployment rate, the organizational changes in the company and the recent deals that it was involved in, the overall political situation and the customers' interest in the products of the company.
3. *Combined approach*: This approach tries to incorporate both types of information available (the historical data as well as the data characterizing the state of the economy) in order to make a prediction.

5.2 Literature Review

The important work related to the prediction of stock markets is discussed in this section. The work done in the field of ARIMA modeling is discussed first. Debadrita Banerjee (2014) used and implemented ARIMA models for the Indian stock exchange. ARIMA of order (1,0,1) was used for forecasting the share price. ARIMA has been commonly used to make predictions in a variety of fields. Jiang H et al. (2021) forecasted cotton and sporting goods prices using ARIMA. As compared to simple exponential smoothing (SES) and holt two parameters, this method produced better prediction accuracy. Fattah et al. (2018) used ARIMA for demand forecasting and discovered that fitting the model involves a large number of observations and estimations from ACF and PACF models. They discovered that the AIC is not the minimum value and that non-linearity exists after using Box Jenkin's method with autocorrelation and partial correlation.

Anurag Barthwal et al. (2021) applied the ARIMA model in the prediction of air quality levels in an urban environment. They concluded that ARIMA was better suited for the prediction of a non-stationary time series than regression-based methods. Temur et al. (2019) used the hybrid model LSTM-ARIMA for predicting housing prices in Turkey, where they pay more attention to the linear aspect of time series and non-linear aspect of LSTM. In their study, they concluded that the approach of using the hybrid ARIMA-LSTM model gives better accuracy. Khandelwal et al. (2015) used a technique known as discrete wavelet transform (DWT) to separate linear and non-linear factors of the time series. They used four datasets for this purpose, to show the clear comparison wherein ARIMA was used for the linear component and ANN for a non-linear component of time series.

Stock movement prediction using ARIMA was applied by A. Ariyo et al. (2014) in New York Stock Exchange and Nigeria Stock Exchange, and they concluded that this model is acceptable for short-term prediction. C. N. Babu et al. (2014) implemented ARIMA and GARCH together and concluded a good amount of accuracy. A. A. Adebiyi, et al. (2014) compared the result obtained by applying ARIMA and neural networks and their conclusion was found to be contradictory over best fitting. R. Majhi et al. (2009) performed market index analysis with bacterial foraging optimization and adaptive BFO strategies and found the performance of these methods convincing. L. Yu et al. (2009) placed the support vector method in contrast with GA based support vector machine. The combined system based on genetic algorithm and SVM was recommended to forecast the market.

Some authors are additionally of the opinion to use PHM (hybrid model) the place the weights had been decided with the aid of Genetic Algorithm (Wang et al., 2012). The price of the stock had been also been forecasted the usage of chaotic mapping, firefly algorithm and assist vector regression (Kazem et al., 2013). Hybrid ARIMA and ANN have been used for the prediction of the stock data (Babu et al., 2014). B. U. Devi et al. (2013) analyzed for this purpose in which instances collection analysis used to be carried out the usage of the ARIMA model for making clever decisions.

5.3 Objectives and Study

The most important goal of this paper is to track the motion of the commodity market. It is one of the indices traded in India by investors who are interested in making investments in commodity indexes without getting into the future market. These values additionally fluctuate comparable to stocks. This index additionally offers exposure and acts as a benchmark.

5.4 Data and Methodology

The data that has been used for this study is the closing price of indices of the commodity index and the duration of the time series data is from the year 2015 to the year 2021. The time series consists of 2,000 observations. This information was analyzed with the help of the ARIMA model to determine how the price will vary in the coming years.

5.5 Data Decomposition

The first step in processing the commodity index time-series data is to find out what fraction of the data possesses trend seasonality if the data-set is seasonal. Data decomposition is an important pre-processing task that is required to be performed on the time series dataset, which is used to determine the trend using past data (Brownlee, 2020). But, there are various effects such as holidays and weekdays, which are applied in limited occurrence. These components can either be multiplied or added in an additive model. As a rule of thumb, a multiplicative model must be selected if the seasonality of the timeseries is increasing.

The diagram above (Figure 5.1) shows that there is a clear and distinguishable stationary in the time series data. Seasonality, trend and periodicity/cyclicity are discussed in the following sub-sections.

5.5.1 Seasonality

It is generally a kind of pattern due to seasons and it is generally said that seasonality depends on human behavior (Brownlee, 2020).

5.5.2 Trend

The trend tracks the direction or movement of the time series. It may either increase or decrease in a year or it is stable. Figure 5.2 shows the seasonality

Figure 5.1 Time-series of closing price of the commodity index for the duration from the year 2015 to the year 2021.

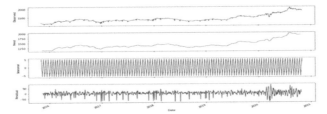

Figure 5.2 The seasonality of the time-series of closing price of the commodity index for the duration from the year 2015 to the year 2021.

for the timeseries of the closing price of the commodity index for the duration from the year 2015 to the year 2021.

5.5.3 Cyclicity

It is a kind of pattern returning every year or at the same time.

5.6 Augmented Dicky Fuller (ADF) Test

When the method of decomposition is applied, the subsequent assessment is to decide whether or not stationarity exists in the timeseries. In order to use ARIMA to its best, stationarity is one of the necessary conditions as it is established that stationarity records have regular implications and variance over time (Aburto et al., 2013). The well-known test to take a determine stationarity on a time series dataset is the ADF. It is no longer advocated to proceed for prediction without checking the stationary of the time series. In order to take a look at stationarity, the cost of p for specific time intervals is: for the entire time sequence, the value of p is taken is $p = 0.99$ and for zero to 3 months, $p = 0.183$. For 3–6 months it is 0.671 and for 6–9 months, it is 0.70. For a time duration of more than 12 months, it is 0.8332.

Before proceeding to generate time series forecasts, it is required that the commodity index time series is stationary: both mathematically as well as visually. If a timeseries exhibits non-stationary, it must be checked for stationarity after differencing. The stationarity is determined with the help of the ADF test. Once the timeseries is found to be stationary, we proceed to generate auto-correlation and partial auto-correlation function plots.

5.6.1 Auto-correlation function (ACF)

Auto correlation is a statistical term that describes the correlation or lag in a time series dataset. This is a key statistic when it comes to times series analysis because whenever we are working with a time series dataset, there is always a question of whether or not the previous observation will influence the recent one. The autocorrelation function is used to determine the correlation on a time scale. The steps on the time scale are called lags. For efficient prediction of a time-series data-set, we always try to find out and test if such correlation is present or not.

The main objective of plotting the autocorrelation function is to decide whether to use auto regressive or moving average models or else to combine the two models, which results in the creation of the ARIMA model. The current commodity index value in the case of an auto regressive time series model depends on the 'p' past values of the commodity index:

$$I_t = f(I_{t-1}, I_{t-2}, I_{t-3}, I_{t-4}, \ldots, I_{t-p}, \phi_t) \tag{5.1}$$

where,

I_t denotes the closing price of the commodity index to be determined, and

$I_{t-1}, I_{t-2}, I_{t-3}, I_{t-4}, \ldots, I_{t-p}$ represent the 'p' past values of the closing price of the commodity index, and

ϕ_t is the random error or Gaussian white noise term.

72 Analysis and Prediction of Closing Price of Commodity Index

In case of the moving averages time series method, for the prediction of the current closing price of commodity index, we use the error terms that follow Gaussian white noise:

$$I_t = f(\phi_{t-1}, \phi_{t-2}, \phi_{t-3}, \phi_{t-4}, \ldots, \phi_{t-q}) \quad (5.2)$$

where,

I_t is the closing price of the commodity index to be determined, and, $\phi_{t-1}, \phi_{t-2}, \phi_{t-3}, \phi_{t-4}, \ldots, \phi_{t-n}$ are the 'q' past random Gaussian white noise errors.

The value of p, q need has to be assigned to be able to use the AR and MA models. So, while applying autocorrelation, if the correlation is positive during the first lag, it will be a good decision to use AR. On the other hand, a negative correlation in the beginning, indicates that the MA will be a better alternative to use. This method helps us to determine the values of p, d and q to be used in the model, where 'd' is the number of times we are required to difference the time series in order to obtain a stationary time-series. The value of 'p' depends on the count of the lag observed and 'd' is decided how many times the information was differenced. The value of 'q' here is the moving average size (Wang, 2012). The correlation between the current and the past values of the commodity index is denoted using ACF and its value ranges from $-1 - +1$ (Figure 5.3).

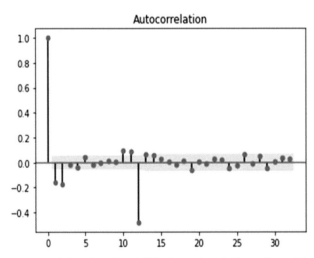

Figure 5.3 Auto-correlation function (ACF) plot against the lag values of the time-series of closing price of the commodity index for the duration from the year 2015 to the year 2021.

5.6 Augmented Dicky Fuller (ADF) Test 73

As shown in Figure 5.3, the peak values of ACF are obtained at lag values of 0, 1, 2 and 12 respectively.

5.6.2 Partial Autocorrelation Function (PACF)

Auto correlation measures the similarity between a time series and a lagged version of itself. However, the coefficients also capture second-hand effects. For instance, if we examine the value of the auto correlation coefficient for the third lag in Figure 5.3, it captures the direct and indirect ways in which the lagged series affects the original one. We refer to all other channels through which past data of the current timeseries is affected. In this specific case, these second-hand effects come in the form of prices. The value of the commodity index three days ago affects the value 2 days ago and the value 1 day ago affects the present prices. If we wish to determine only the direct relationship between the time series and its lagged version, we need to compute the partial auto correlation to do this. It is denoted by PACF (Banerjee, 2014).

5.6.3 ARIMA Model

ARIMA model has three orders: p, d and q. The orders p and q symbolize the AR and MA lags. With the ARMA models (Tamatta, 2018), the order 'd' is the integration order. It represents the quantity of instances we want

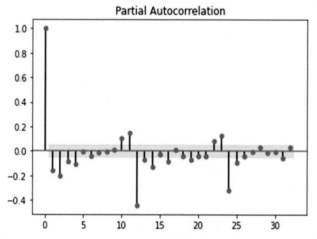

Figure 5.4 Partial auto-correlation function (PACF) plot against the lag values of the time-series of closing price of the commodity index for the duration from the year 2015 to the year 2021.

to combine from the commodity index time sequence to achieve stationarity (Rahimi and Khashei, 2018). Convention dictates that we constantly enter the three orders in a repetitive manner: first p, then d and eventually q. The value of p represents the AR elements, d the built-in ones and q the MA elements. The p order AR elements and q order MA elements are derived from the commodity index time sequence data.

The ARIMA model has two categories: any model of ARIMA($p,0,q$) is = ARMA(p,q) model as we are not including any degree of changes or differencing. Similarly,

ARIMA($0,0,q$)=MA(q), and

ARIMA($p,0,0$)=AR(p).

This is how models are connected. The equation of ARIMA model for prediction of closing price of commodity index has all orders as 1, ARIMA(1,1,1).

The entire ARIMA (p,d,q) model is nothing more than ARMA (p,q) model for a newly generated time series that is stationary. The seasonal ARIMA is used for seasonal prediction and it uses a multiplicative model (Hyndman et al., 2020). In our study, we applied non-seasonal ARIMA. Non-seasonal ARIMA model is obtained when we use auto regression and moving average models in conjunction, with differencing. This model is generally denoted by ARIMA (p,d,q) where p represents the auto-regression part, d is for a number of times the differencing of the index price time-series is performed, and q represents the moving average part and its order, which is generally the dependency between the error and observed values (Kazem, 2013). In this study, non-seasonal ARIMA is used to obtain the results.

5.7 Results and Analysis

The time series data for the duration from 01-12-2015–30-11-2020 is used to develop the proposed model. Figure 5.5 shows the variation in the time-sequence of the closing price of commodity index from 01-12-2015–30-11-2020.

The predictions obtained with the help of the ARIMA(1,1,1) model are graphically displayed in Figure 5.6. The time sequence data of the closing price of commodity index that is used for building the ARIMA model is represented by blue line and the forecast for the forthcoming days is represented with the help of red line.

5.7 Results and Analysis

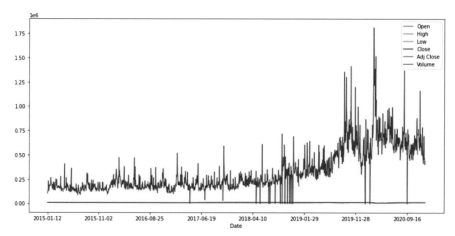

Figure 5.5 Time-series of closing price of the commodity index for the duration from 01-12-2015–30-11-2020.

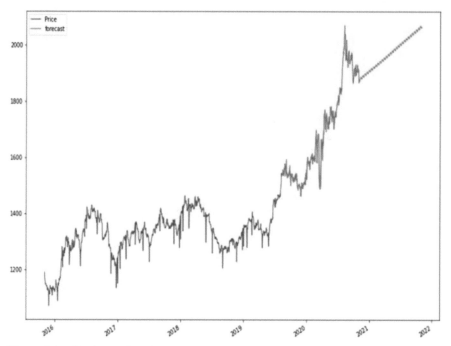

Figure 5.6 The time-series of closing price of the commodity index for the duration between the year 2015 and the year 2021 (Blue line represents previous time-series values used for building ARIMA model and red line represents the forecast generated by the model).

The actual commodity price, forecasted price, absolute error and error percent with the application of the proposed ARIMA(1,1,1) model for the week starting from 19-12-2020–27-12-2020 are detailed in Table 5.1.

Figure 5.7 shows the difference between the actual and the predicted closing prices of the commodity index for the test duration from 19-12-2020–27-12-2020.

Table 5.1 The actual commodity price, forecasted price, absolute error and error percent for the time duration from 19-12-2020–27-12-2020

Day	Actual Price	Predicted Price	Absolute Error	Error %
19-12-2020	1911.7	1905.9	5.80	0.30
20-12-2020	1915.2	1909.7	5.50	0.29
21-12-2020	1904.1	1916.7	12.60	−0.66
22-12-2020	1905.8	1930.1	24.30	−1.28
23-12-2020	1905.3	1910.2	4.90	−0.26
24-12-2020	1911.2	1906.1	5.10	0.27
25-12-2020	1879	1901.3	22.30	−1.19
26-12-2020	1868.5	1913.4	44.90	−2.40
27-12-2020	1879.1	1981.5	62.40	−3.32

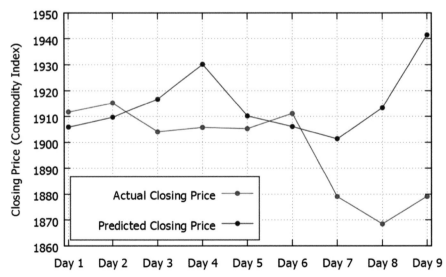

Figure 5.7 The difference between actual and the predicted closing prices of the commodity index for the test duration between 19-12-2020–12-2020 (Green line represents the actual closing price and maroon line represents the forecast generated by the model).

The average absolute error for the proposed model is 20.87 and the average absolute percentage error is 1.10%. The ARIMA model is found to have a mean absolute percentage error (MAPE) value of 1.11% and a root mean square (RMS) value of 28.42, which is within an acceptable range. Hence, the proposed method is found to be able to predict commodity index prices with high accuracy and low error rates.

5.8 Conclusion and Future Work

The present study is related to the ARIMA model which was applied on time-series data of commodity index. From the results obtained, we observe that the index prices closely resemble the actual prices on the give dates. In this way, we can conclude that the ARIMA model was successfully implemented on the timesseries data and it is capable of yielding good results with high accuracy. The ARIMA(1,1,1) model performed well with an average absolute error of 20.87 and an average absolute percentage error of 1.10%. It exhibited high accuracy and low error rates. The MAPE and RMS values for the proposed model were 1.11% and 28.42 respectively. Hence, it is concluded that the ARIMA method is most suitable for the short-term prediction of the closing price of a commodity index.

In the future, we would like to optimize this model to be used on different indices. As described above, in the paper, we have analysed the data of the time series in the stock market using the ARIMA model. As shown in the results, the predicted price using the ARIMA model is very close to the actual price on the given dates. This indicates the successful implementation of the ARIMA model on the time series data. In the future, we hope to analyze the dynamic data of the stock market. We propose an optimized hybrid ARIMA model for the dynamic data in the future.

References

Aburto, L., Weber, R. (2007), Improved supply chain management on hybrid demand forecasts. *Applied Soft Computing*, 7(1), 136–144.

Adebiyi. A. A., Adewumi, A. O. and Ayo,C. K. (2014). Comparison of ARIMA and artificial neural networks models for stock price prediction. *Journal of Applied Mathematics*, 2014.

Ariyo, A. A., Adewumi, A. O., Ayo, C. K. (2014, March). Stock price prediction using the ARIMA model. In 2014 *UKSim-AMSS 16th International Conference on Computer Modelling and Simulation*, 106–112, IEEE.

Babu, C. N., and Reddy, B. E. (2014). Selected Indian stock predictions using a hybrid ARIMA-GARCH model. In 2014 *International Conference on Advances in Electronics Computers and Communications*, 1–6, IEEE.

Babu, C. N., Reddy, B. E. (2012). Predictive data processing on average global temperature using variants of ARIMA models. *IEEE-International Conference On Advances In Engineering, Science And Management* (ICAESM-2012), 256–260.

Babu, C. N., Reddy, B. E. (2014). A moving-average filter based hybrid ARIMA–ANN model for forecasting time series data. *Applied Soft Computing*, 23, 27–38.

Banerjee, D. (2014). Forecasting of Indian stock market using time-series ARIMA model. In 2014 2nd *International Conference on Business and Information Management* (ICBIM), 131–135, IEEE.

Barthwal, A. and Acharya, D. (2021). An IoT based sensing system for modeling and forecasting Urban Air Quality, *Wireless Personal Communications*, Vol. 116, Pages 3503–3526, Doi: https://doi.org/10.1007/s11277-020-07862-6

Brownlee, J. (2021). A gentle introduction to autocorrelation and partial autocorrelation, Retrieved from https://machinelearningmastery.com

Chen, S. (2020). Stock Market. Retrieved from www.investopedia.com

Devi, B. U., Sunda,r D., Alli, P. (2013). An effective timeseries analysis for static trend prediction using ARIMA model for nifty midcap-50.*International Journal of Data Mining & Knowledge Management Process*, 3(1), 65.

Fattah, J., Ezzine, L., Aman, Z., Moussami, H. E., Lachhab, A. (2018). Forecasting of demand using ARIMA model. *International Journal of Engineering Business Management*, 10:184797901880867.

Hasin, M. A .A., Ghosh, S., Shareef, M. A. (2011). An ANN approach to demand forecasting in retail trade in Bangladesh. *International Journal of Trade, Economics and Finance*, 154–160.

Hyndman, R. J. et al. (2020), Seasonal ARIMA models, Retrieved from https://online.stat.psu.edu/

Hyndman, R. J. et al. (2021), Non-seasonal ARIMA models, Retrieved from https://otexts.com

Jiang, H., Leng. X., Choi, M. (2021, Feb). Computer prediction of cotton and sporting goods prices in e-commerce environment based on ARIMA model. *The International Journal of Electrical Engineering & Education*. Doi:10.1177/0020720920983706

Kazem, A., Sharifi, E., Hussain, F. K., Saberi, M., Hussain, O. K. (2013). Support vector regression with chaos-based firefly algorithm for stock market price forecasting. *Applied Soft Computing*, 13(2), 947–958.

Khandelwal, I., Adhikari, R., Verma, G. (2015). Time series forecasting Using hybrid ARIMA and ANN Models Based on DWT Decomposition. *Procedia Computer Science*, 48, 173–9.

Majhi, R., Panda, G, Majhi, B Sahoo G. (2009). Efficient prediction of stock market indices using adaptive bacterial foraging optimization (ABFO) and BFO based techniques. *Expert Systems with Applications*, 36(6), 10097–10104.

Mehdi Askari and Hadi Askari (2011). Time series grey system prediction-based models: Gold Price Forecasting. *Trends in Applied Sciences Research*, 6, 1287–1292. Doi: 10.3923/tasr.2011.1287.1292

Pai, P. F. and Lin, C. S. (2005). A hybrid ARIMA and support vector machines model available in stock price forecasting. Omega, 33(6), 497-505. Doi: 10.1016/j.omega.2004.07.024

Stephanie (2020), Augmented Dickey Fuller Test, Retrieved from www.statisticshowto.datasciencecentral.com

Tamatta, R. K. (2018). Time series forecasting of hospital Inpatients and Day case waiting list using ARIMA, TBATS and Neural Network Models (Doctoral dissertation, Dublin, National College of Ireland)

Temür, A. S., Akgün. M., Temür. G. (2009). Predicting housing sales in Turkey using Arima, Lstm and Hybrid models. *Journal of Business Economics and Management*. 20(5), 920–38. Doi: 10.3846/jbem.2019.10190

Wang, J. J., Wang, J. Z., Zhang, Z. G., Guo, S. P. (2012). Stock index forecasting based on a hybrid model. *Omega*, 40(6), 758–766.

Yanovitzky, I., Van Lear, A. (2008). Time series analysis: Traditional and contemporary approaches. In A. F. Hayes, M. D. Slater, & L. B. Snyder (Eds.), *The SAGE Sourcebook of Advanced Data Analysis Methods for Communications Research*, Thousand Oaks, CA: Sage Publications, 89–124.

Yu. L., Wang, S., Lai, K. K. (2005). Mining stock exchange tendency using GA-based support vector machines. *International Workshop on Internet and Network Economics*, 336–345. Springer, Berlin, Heidelberg.

6

Neural Networks Analysis of Suspended Sediment Transport Time Series Modeling in a River System

M. Harini Reddy[1],*, N. Manikumari[2],*, M. Mohan Raju[3],*, Dinesh C. S. Bisht[4], A. Naresh[5], Harish Gupta[6], and M. Gopal Naik[7]

[1]Department of Civil Engineering, Annamalai University, Annamalai Nagar-608002, India
[2]Professor, Department of Civil Engineering, Annamalai University, Annamalai Nagar-608002, India
[3]Assistant Executive Engineer, Nagarjunasagar Project, Irrigation & CAD (Projects Wing) Department, Govt. of Telangana State,
Hill Colony-508202, India
[4]Department of Mathematics, Jaypee Institute of Information Technology, Noida - 201304, India
[5]Department of Civil Engineering, Osmania University, Hyderabad-500007, India
[6]Department of Civil Engineering, Osmania University, Hyderabad-500007, India
[7]Department of Civil Engineering, Osmania University, Hyderabad-500007, India
E-mail: harinireddy589@gmail.com; nmanikumariau@gmail.com; mmraju.swce@gmail.com; drbisht.math@gmail.com; ayyaure@gmail.com
*Corresponding Authors

Abstract

The present study is a real-time application of Artificial Neural Networks (ANNs) to estimate, predict and forecast the suspended sediment transport in streams and river systems using time series data. NN approach is a crucial and readily adaptable important methodology when hydrograph/unit hydrograph methodology and conventional mathematics are inconvenient to apply for an emergency situation to act as the sediment runoff time series data is highly complex in nature i.e. situations like availability of less amount of data, sometimes the very high and huge quantity of data, poor quality of data, erratic nature of data, need of emergency prediction and forecasting, etc. The present study is pertaining to a tributary; namely *Peddavagu* in the Godavari river system in India. A back propagation algorithm was employed in the methodology among the layers of multi-layered feed-forward neural network with one hidden layer and two hidden layer approaches. The study also demonstrates the beauty of an application of computational intelligence/ artificial intelligence technique as how it is superior to the conventional mathematical modeling. Sediment rating curve (SRC), mathematical modeling was carried out to present a comparative performance of the applied methodology. Statistical performance evaluation criteria using root mean square error (RMSE), correlation coefficient (CC) and coefficient of determination (DC) comprehend the sensitivity in applying the ANNs for complex suspended sediment time series to predict and forecast the sediment hydrology of river systems. Hysteresis effect of sediment runoff is also a component in the study to reveal a result of many number of sediment concentration magnitudes for a specific runoff magnitude in the river flow system.

Keywords: suspended sediment, runoff, time series, neural network, modeling, forecasting, back propagation algorithm, rating curve, hydrograph and hysteresis.

6.1 Introduction

Soil erosion, transportation and sedimentation is a major complex hydrological phenomenon of suspended sediment in the runoff process of a river system and which provides an objective function of typical estimation and modeling of sediment transportation time series (Onstad and Foster, 1975). ANNs are the best artificial intelligence methodology to handle and simulate such complex, the non-linear hydrologic processes for estimation (ASCE, 2000a and ASCE, 2000b), as the process shows very highly significant spatial and temporal variations. The chapter under study is intended to examine and comprehend the fluvial scenario of the *Peddavag*u river system, which is a tributary system of the Pranhita sub-basin of the Govdavari basin in the southern Indian peninsula with a drainage basin area of about 4260 km^2.

The prime objective of the chapter is to model the suspended sediment transport time series of the *Peddavagu* river system located in a very heavy rainfall zone, as how the river system affects the downstream with huge carrying sediment loads. A sediment runoff relation for a river system was developed and modeled using ANN with the data recorded by the hydrological gauging station of the *Peddavagu* river and SRC analysis has also been carried out as a comparative study to present the very fine performance of the ANN methodology when an emergent and urgent assessment/estimation of sediment runoff is required. ANN methodology aimed to develop runoff-sediment relation using the time series data of runoff and suspended sediment concentration values as inputs with a specified time step. ANNs with back propagation training algorithm applied to model the suspended sediment transport time series with the following objective functions:

1. ANN and SRC models development for Peddavagu tributary system of Pranhita sub-basin of Godavari river system.
2. Validation of developed models
3. Performance evaluation of formulated models for Peddavagu tributary system.

6.2 Artificial Neural Networks

ANN is typically considered a universal approximator, as the ANNs are capable of developing a relationship between input and output patterns such as complex non-linear modeling, pattern recognition, control, etc. ANN information processing system evolved on the basis of the human brain biological neural network (Haykin, 1994) and generalized by neuro-biology mathematical models. Most commonly ANNs are appeared to be the same as regression-based models in engineering hydrology. ANN architecture (Figure 6.1) framework is a subject of defining the objective function, as how the output is estimated or targeted with the pre-determined number of inputs to solve the objective function along with the number of hidden layers and hidden nodes in the intermediated layers that are trialed for least mean squared error (MSE). ANN architecture determination and optimization of a number of nodes in the hidden layers is generally assessed by the trial-and-error procedure. In the current study, one hidden and two hidden layers architectures employed using the number of hidden nodes in the hidden layers was tried from equal to and doubled number of nodes present in the input layer. Input layers receive the data from the outside world and the

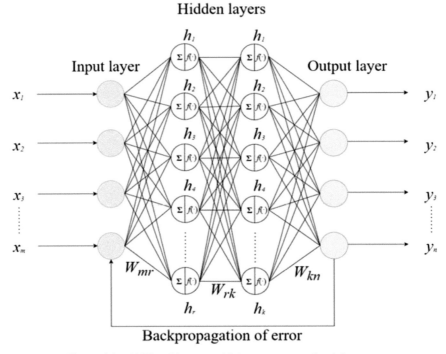

Figure 6.1 ANN architecture with input, output and weight vectors.

data or signals are transmitted through hidden layers undergo the process of optimization to result in significant targets at the output layer, finally the estimated outputs are delivered to the outside world through output layers. When the signals are transmitted from one layer to the other each individual signal is multiplied by its individual connection strength or weight; later the weighted signals are summed up at the next stage neuron to feed the net value to the activation function i.e. sigmoidal transfer function before the transformed signal is fed to the next layer towards the output layer.

Normalization is the process of filtering and scaling the times series data to bring to a uniform range [0,1] to reduce the outlier effect because low and high-frequency data signals may alter the regular path of the main signal, hence the time series properly normalized to work with the neural network (Karunanithi et al., 1994). Further, neural networks are capable of doing filtering themselves during data processing in advance in the transmission of signals through the network. Back propagation algorithm is the most widely used algorithm in water resources research applications usually employed among the systematically trained neural network based on the error-correction rule in order to update connecting strengths (Kisi, 2007). Randomly connecting strengths are initialized (between -1.0 and 1.0 or $-0.5 - +0.5$) during the iterative training process in the BP algorithm and are updated during feed-forward calculations in every iteration. Updation of connecting strengths is determined by learning rate (η) and connecting strength modifications are proportional to the negative gradient of the error results in the iterative process. Selecting a bigger value for 'η' results in rapid learning and for lower values slow learning thereby unusual oscillations may occur in weights updation, hence an optimum value for 'η' should be chosen in the training process of the network. In many hydrologic engineering problems, the value for 'η' is considered in the range between 0.1 and 0.9 (ASCE, 2000a and ASCE, 2000b). Momentum rate 'α' is another factor in gradient descent method incorporated to control the effect of instability in the training process (Sato, 1991) and mostly a value for 'α' is 0.9 chosen in water resources research problems that it exists in closed interval from 0–1 i.e. [0,1]. The compacted and smallest possible neural network architecture is selected based on Akaike's Information Criterion (AIC) and it works on the principle of minimal RMSE, as AIC trades for a compact NN architecture with lower values to work out (Raju, 2008 and Raju, 2011; Bisht et al., 2009 and Bisht et al., 2010). The reader of this chapter is requested to go through the literature on ANNs and their applications in Karnin, 1990; Hario and Jokinen, 1991; Refenes and Vithlani, 1991; Sato, 1991; Nadi, 1991; Zurada M. Zacek 1992; Kohonen, 1998 and the other related references to understand the subject of ANN approach of solving objective function through soft computing in a better way, as the chapter is mainly intended to present the prime results of the suspended sediment transport modeling and estimation using neural networks.

6.3 Hydrological Study Area

The river *Peddavagu*, a tributary of *Panhita river*, rises from the hills near Ghatpalli, Mathira Thanda villages in Godavari basin and while traversing for about 161 km in Utnur, Asifabad and Sirpur Taluqs of erstwhile Adilabad district of Telangana State in India (Figure 6.2) drains an area of 4260 km^2 and joins the Parent river near Rampur village. The river flows bounded by hill ranges and thick forest up to about 65 km near Ada village, where the catchment area is 1132 km^2. There after the ridges gradually widen out. The catchment area of the river lies in a heavy rainfall zone of the Adilabad district

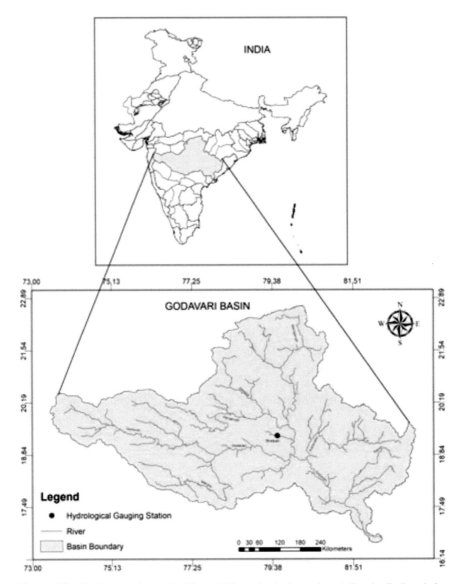

Figure 6.2 Peddavagu river with Bhatpalli hydrological gauging station in Godavari river system.

of Telangana State where the average annual rainfall is about 1050 mm. The area receives more than 90% of the rain during the South-West monsoon (June–October) and consequently, total inflow is received between June and

October only. High floods are expected, during August and September and thereafter the flow in the river diminishes down considerably. The Peddavagu river is having its own catchment comprising of thick forest within it and with tributaries Bapur vagu (145 km^2), Wankidi vagu (363 km^2) and Yerra vagu (109 km^2) on its left flank and Tiryani vagu (1075 km^2) on the right flank. The dependable annual average yield at 75% probability worked out to be about 15 TMC by the Central Water Commission (CWC) in India.

As the catchment covered by mostly hilly regions with the thickly wooded forest is usually received by server floods in the past and to make a note during the year 1958 in the month of August the catchment was received by 270 mm rainfall in one day and 35mm on next day brought a disaster to the land, life, agriculture and other properties in the downstream of the catchment known to be the highest rainfall ever recorded in the history of Peddavagu river in last 200years. The floods of such a magnitude resulted from a catchment area of nearly about 4500 km^2 became a serious concern to frame hydrological studies to estimate and forecast runoff and sediment transportation in the Peddavagu river system. By considering similar typical hydrological events in the study basin as a triggering point the study chapter was framed to assess the suspended sediment runoff estimation and forecasting for the Peddavagu tributary system as it contributes heavy runoff and sediments loads to the Godavari basin along with the Pranhita sub-basin in southern peninsular India. The study is mainly intended to comprehend the quick and sudden watershed responses of intermittent kinds of river systems into large river basin in respect of its sediment transportation and runoff contributions.

6.4 Methodology

The study laid its foundation on the proven hydrological system analyses carried out by ANN methodology such as rainfall-runoff modeling, runoff-sediment modeling, river stage estimation and its forecasting, groundwater modeling, hydrological time series, etc. The study adopted a multi-layered feed-forward neural network with one hidden layer and two hidden layers architectures applied by back propagation algorithm and sigmoid activation transfer function. The hydrological daily data of runoff and sediment concentration procured for the *Bhatpalli* hydrological gauging station on the river *Peddavagu* for 20 years from June 1996–May 2016 in coordination with the Central Water Commission, New Delhi, India. As per the workspace requirement by the application of a valid software used in the study, 75% of the data was used for training, 15% of the data were used for testing and

15% of the data used for validation. Modeling the output as the estimation of sediment concentration designated at a time step 't', i.e. C_t and has been cited in literature by many authors (Cigizoglu, 2002; Tayfur and Guldal, 2005; Sarkar et al., 2010; Naresh et al., 2019) that the current sediment concentration can be mapped principally by considering the current value of runoff (discharge); additionally, the sediment concentration is mapped by the runoff and sediment concentration at the previous times along with current runoff/discharge values (Govindaraju and Kavvas, 1991; Wu et al., 1993). Therefore, in addition to Q_t, i.e. discharge at time step 't', other variables such as Q_{t-1}, Q_{t-2} and C_{t-1}, C_{t-2} were also considered in the input (Nagy et al., 2003; Jothiprakash and Garg, 2009; Nourani et al., 2021). The input and output data combinations considered to train the neural network for the study have been shown the Table 6.1 below. Whereas, in the design of sediment rating curve analysis, the input-output variables used for ANN-1 were only used in modeling, as the variables are prime and fundamental variables in the development of the models to start with. 10 models were designed to execute the objective function to implement various numbers of trials with one and two hidden layer methodology for a particular model and the best representative models were selected for presentation of multilayer neural network and ultimately a best-suited model was drawn from two methodologies to represent the gauging station under study. From there the time series fluvial scenario of the hydrological gauging station under study is analyzed based on a representative ANN model.

Table 6.1 Runoff-sediment models developed for the study

Model	Hidden Layers Applied	Output	Inputs
ANN – 1	One	C_t	Q_t
ANN – 2	One	C_t	Q_t, Q_{t-1}, C_{t-1}
ANN – 3	One	C_t	$Q_t, Q_{t-1}, Q_{t-2}, C_{t-1}, C_{t-2}$
ANN – 4	One	C_t	$Q_t, Q_{t-1}, Q_{t-2}, Q_{t-3}, C_{t-1}, C_{t-2}, C_{t-3},$
ANN – 5	One	C_t	$Q_t, Q_{t-1}, Q_{t-2}, Q_{t-3}, Q_{t-4}, C_{t-1}, C_{t-2}, C_{t-3}, C_{t-4}$
ANN – 6	Two	C_t	Q_t
ANN – 7	Two	C_t	Q_t, Q_{t-1}, C_{t-1}
ANN – 8	Two	C_t	$Q_t, Q_{t-1}, Q_{t-2}, C_{t-1}, C_{t-2}$
ANN – 9	Two	C_t	$Q_t, Q_{t-1}, Q_{t-2}, Q_{t-3}, C_{t-1}, C_{t-2}, C_{t-3},$
ANN – 10	Two	C_t	$Q_t, Q_{t-1}, Q_{t-2}, Q_{t-3}, Q_{t-4}, C_{t-1}, C_{t-2}, C_{t-3}, C_{t-4}$

Where Q=Runoff in cumec C=Suspended sediment Concentration in g/l

6.4.1 Mathematics of SRC

SRC for the estimation of sediment concentration runoff transportation in a river system is a mathematical relation between sediment concentration and runoff in a stream or river having the data applied for the functional relation with time lags such as daily, ten daily, weekly, monthly, seasonally, annually, etc. for study under consideration (Ferguson, 1986; McBean and Al-Nassri, 1998). Mathematical construction of an SRC by a logarithmic transformation of data to determine the best fit line using a linear least square regression as:

$$C = aQ^b \qquad (6.1)$$

Logarithmic transformation on log-log paper represents straight line

$$logC = loga + blog(Q) \qquad (6.2)$$

C = sediment runoff (concentration or load), Q = runoff or discharge and a & b are regression constants.

Using the sediment rating curve model (6.2) the sediment rating equations computed for *Peddavagu* river for *Bhatpalli* hydrological gauging station as below:

$$C = 4.980E - 04\ Q^{0.790} \qquad (6.3)$$

The performance evaluation was executed with the statistics parameters viz. RMSE, correlation coefficient (R) and coefficient of efficiency (CE) or DC. RMSE evaluates the residual error (Yu and Lee, 1994) and DC is a measure of comparative relative performance between initial variance and standardization of residual variance (Nash and Sutcliffe, 1970). The determination coefficient is synonymously referred to as a coefficient of efficiency and represents the fraction of variance that is explained by the regression. The value closer to unity indicates the better statistical performance of the model under evaluation w.r.t DC (Haan, 1977). The reader is herewith suggested to refer to the mathematical formulae for the above presented statistical parameters in the cited references and from standard mathematics textbooks as well.

6.5 Results

The statistical analysis of ANN models and SRC model described numerically in Table 6.2 for the hydrological gauging station under study and the best-performed models were compared with the SRC method to show the

Table 6.2 Performance comparison of ANN modeling and SRC method for Peddavagu river at Bhatpalli hydrological gauging station.

ANN Model	Architecture	Training			Testing		Validation	
		RMSE	R	DC	R	DC	R	DC
ANN1	[1-2-1]	0.065	0.982	0.957	0.907	0.743	0.912	0.667
ANN2	[1-2-1]	0.557	0.986	0.966	0.951	0.856	0.946	0.805
ANN3	[5-6-1]	0.058	0.989	0.969	0.949	0.850	0.945	0.794
ANN4	**[7-5-1]**	**0.054**	**0.992**	**0.974**	**0.979**	**0.918**	**0.983**	**0.956**
ANN5	[9-11-1]	0.059	0.987	0.969	0.944	0.852	0.940	0.819
ANN6	[1-3-5-1]	0.061	0.981	0.965	0.923	0.830	0.917	0.695
ANN7	[3-5-5-1]	0.060	0.983	0.961	0.957	0.877	0.953	0.871
ANN8	[5-3-5-1]	0.067	0.979	0.963	0.961	0.889	0.930	0.865
ANN9	**[7-5-6-1]**	**0.059**	**0.988**	**0.967**	**0.967**	**0.907**	**0.972**	**0.917**
ANN10	[9-7-5-1]	0.063	0.989	0.975	0.955	0.861	0.958	0.888
SRC	–	0.512	0.929	0.690	0.890	0.880	0.877	0.771

superior and high accurate modeling process. During the presentation of results in Table 6.2 ANN4 was chosen as the best representative model for the study area and RMSE values of SRCs are highly deviate in their magnitude; performed very poorly when compared with respective ANN models. In the statistic of correlation coefficient all most all the ANN models performed (more than 0.907 i.e. 91%) so well and the values are very high in some cases i.e. more than 99%, further, all the ANN models have shown good generalization capability in the training phase, testing phase and validation phase when compared with SRC conventional method. The developmental increment or the improvement in 'R' statistic in the validation phase reveals the capability of the model for good generalization. The performance of the SRC model is very normal in 'R' statistic and significantly lesser efficient in three criteria as presented, because the calculated sediment values from sediment rating curve model follow a general trend like observed values series thereby higher magnitude R values may results, however there is the significant numerical deviation in observed and estimated sediment concentration, hence the values of RMSE and DC are inefficient for the SRC under study. The depicted representative Figures 6.3 and 6.4 for a best performed ANN model i.e ANN4 for *Bhatpalli* gauging station of Peddavagu river will give a good understanding of the graphical representations of observed and estimated sediment concentration and they are scatter plots in training (a), testing (b) and validation (c) phases.

In the Figure 6.3 a very little amount of mismatch may be seen between the targeted and calculated suspended sediment concentration series during training, testing and validation and in the same lines of presentation the scatter

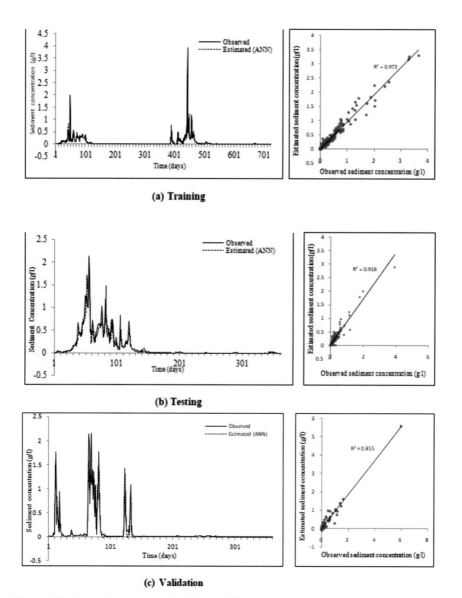

Figure 6.3 Real time series and ANN modelled suspended sediment plots of Peddavagu river at Bhatpalli hydrological gauging station.

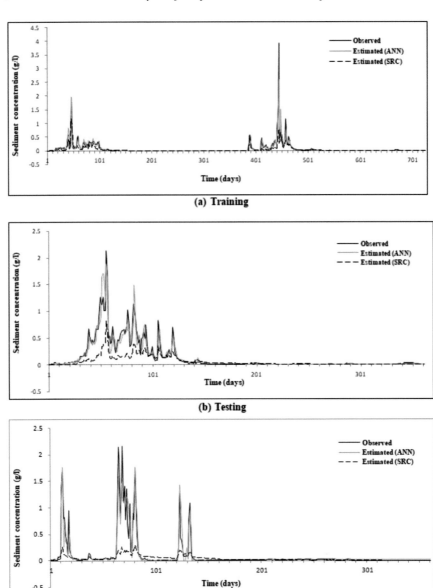

Figure 6.4 Real time series, ANN model time series and SRC time series for Peddavagu river at Bhatpalli.

plots demonstrate the very good correlation values and very negligible or insignificant deviation from the ideal line; otherwise, the deviation from the ideal line may cause in systematic errors. It can be seen from Figure 6.3 the scatter plots in the training phase, testing phase and validation phase show very high values of correlation and the line follow the ideal line. The comparative plots of temporal variation among observed time series, ANN estimated series and SRC estimated series are graphically shown in Figure 6.4. ANN methodology perfectly simulated the trend of observed time series and yielded very fine closer values, whereas SRC couldn't follow the trend of real-time series and seen a high mismatch, which is a synonymous result observed in literature by Kisi, 2007; Sarkar et al., 2010; and Palu and Julien, 2019).

6.6 Hysteresis of Sediment Transport Process

The neural network solution is capable to capture the natural relationship between the runoff and sediment concentration as similar as the observed real-time series in rising and falling limbs of the hydrograph. The real-time hydrograph depicts that various runoff magnitudes may have many number of sediment concentrations in the entire hydrograph; the reason for the variation is that the soil erosion, sedimentation and transportation is high during monsoon season and later diminishes drastically due to settlement in the river system. Such types of spatial and temporal variations in the sediment concentration are best mapped by ANN rather than the conventional methodologies such as linear and non-linear regression mathematical models as shown in Figure 6.5. Hydrological topologies and trajectories are thereafter used to develop a dedicated model that is able to switch between different structural processes and mechanisms and thus offer a suitable construct for the incorporation of trigger events. The presentation of neural network analysis in the current chapter will address a keynote in the precise estimation of events of hydrologic and hydraulic processes involved by hysteresis effects during sediment transportation. The data obtained by ANNs i.e. simulated and estimated sediment concentration using ANNs and the data calculated by SRC for the three phases viz. training, testing and validation compared in the current study with the corresponding real-time series data of sediment concentration and shown graphically below. The hysteresis estimation by ANNs coincides graphically with the hysteresis of observed real-time series data whereas; the SRC is unable to simulate the hysteresis effect with the observed sediment transport process.

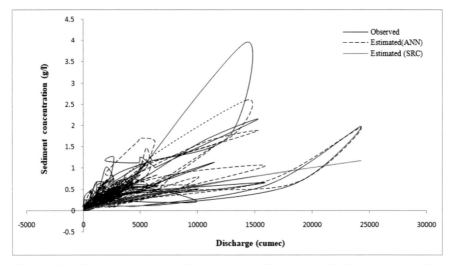

Figure 6.5 Observed and estimated hysteresis of sediment load in Peddavagu river basin.

6.7 Conclusions

The present chapter is a peculiar design of neural network analysis for a sediment runoff time series and to present the superior modeling capability of neural networks over conventional mathematical methods, as the mathematics of sediments, series needs a huge bulk amount of data to dump into a model. The strict guidelines of ASCE 2000a and ASCE, 2000b were executed to solve the prime objective of the study. The current chapter will be a keynote address for a beginner, as how to select the inputs and output parameters along with dependent and independent variable parameters while working with the simulation process. Further to the study, the study also reveals that the hysteresis process can only be described with the help of soft computing as the conventional method is unable to explain the process. At this juncture, it is to mention and acknowledge the facilitating working tool that the tool used for the current research i.e. trial version software available online is alyuda neuroIntelligence.

6.8 Acknowledgements

It is the authors' proud privilege to acknowledge the help and support rendered by the CWC, New Delhi, National Institute of Hydrology (NIH), Roorkee, Department of Soil and Water Conservation Engineering and Department of Irrigation and Drainage Engineering at The G.B. Pant University of Agriculture and Technology, Pantnagar, Department of Civil Engineering, Osmania University, Hyderabad and Irrigation and Command Area Development Department, Government of Telangana State in India in the hydrological time series data acquisition, analyses and providing their guidelines. The authors also mention and acknowledge the facilitating working tool used for the current study i.e. trial version software alyuda neuroIntelligence available online.

References

ASCE Task Committee. on Application of Artificial Neural Networks in hydrology (2000 a). Artificial neural networks in hydrology. I: Preliminary concepts. *Journal of Hydrologic Engineering, ASCE,* 5(2), 115–123.

ASCE, Task committee on Application of Artificial Neural Networks in Hydrology (2000 b). Artificial neural networks in hydrology. II: Hydrologic applications. *Journal of Hydrologic Engineering, ASCE,* 5(2), 124–137.

Bisht, D. C. S., Raju, M. M., and Joshi, M. C. (2009). Simulation of water table elevation fluctuation using Fuzzy-Logic and ANFIS. *Computer Modelling and New Technologies,* 13(2), 16–23.

Bisht, D. C. S., Raju, M. M., and Joshi, M. C. (2010). ANN based river stage – discharge modelling for Godavari river, India. *Computer Modelling and New Technologies,* 14(3), 48–62.

Cigizoglu, H. K. (2002). Suspended sediment estimation and forecasting using artificial neural networks. *Turkish J. Eng. Env. Sci,* 15–25.

Ferguson, R. I. (1986). River loads underestimated by rating curves. *Water Resources Research*, 22(1), 74–76.

Govindaraju, R. S., and Kavvas, M. L. (1991). Modeling the erosion process over steep slopes: Approximate analytical solutions. *Journal of Hydrology*, 127(1-4), 279–305.

Haan, C.T. (1977). Statistical Methods in Hydrology. The Iowa State University Press, Ames.

Hario, H., and Jokinen, P. (1991). Increasing the learning speed of back-propagation algorithm by linearization. Artificial Neural Networks, T. Kohonen et al., 629–634.

Haykin, S. (1994). Neural Networks - A Comprehensive Foundation. Mcmillan, New York.

Jothiprakash, V., and Garg, V. (2009). Reservoir sedimentation estimation using artificial neural network. *Journal of Hydrologic Engineering, ASCE,* 14(9), 1035–1040.

Karnin, E. D. (1990). A simple procedure for pruning back propagation trained neural networks. IEEE Tran. Neural Networks, 1, 239–242.

Karunanithi, N., Grenney, W. J., Whitley, D., and Bovee, K. (1994). Neural networks for river flow prediction. *Journal of Computing in Civil Engineering, ASCE.* 8(2), 201–220.

Kisi, O. (2007). Streamflow forecasting using different artificial neural network algorithms. *Journal of Hydrologic Engineering, ASCE.,* 12(5), 532–539.

Kohonen, T. (1998). Self organization and associative memory. 2nd Ed., Springer Verlag, New York.

McBean, E. A., and Al-Nassri, S. (1998). Uncertainity in suspended sediment transport curves. *Journal of Hydrologic Engineering, ASCE,* 114(1), 63–74.

Nadi, F. (1991). Topological design of modular neural networks. Artificial Neural Networks, T. Kohonen et al., 213–217.

Nagy, H. M., Watanabe, K., and Hirano, M. (2003). Prediction of sediment load concentration in rivers using artificial neural network model. *Journal of Hydraulic Engineering*, ASCE, 128(6), 588–595.

Nourani. V., Gokcekus. H., and Gelete, G. (2021). Estimation of suspended sediment load using artificial intelligence-based ensemble model. *Hindawi Complexity Volume 2021, Article ID 6633760,* 19 pages, https://doi.org/10.1155/2021/6633760

Naresh, A., Raju, M.M., Naik, M.G., Gupta, H., Kumar, A., and Sarkar, A. (2019, December). Sediment transport modeling and hysteresis study for pranahita sub-basin of godavari river system in india. In Hydro-2019 conference proceedings. 2019 *International conference on Hydraulics, Water Resources & Coastal Engineering,* Hydro-2019. Osmania University, Hyderabad, India.

Nash, J.E., and Sutcliffe, J.V. (1970). River flow forecasting through conceptual models. Part 1 – A: Discussion principles. *Journal of Hydrology,* 10, 282–290.

Onstad, C. A., and Foster, G. R. (1975). Erosion modeling simulation for the process of soil erosion by water. *Transaction of the ASCE*, 18(2), 288–292.

Palu, C. M., and Julien, Y. P. (2019) Modeling the sediment load of the Doce River after the Fundão Tailings Dam Collapse, Brazil. *Journal of Hydraulic Engineering, ASCE,* 145(5), 05019002

Raju, M. M. (2008). Runoff – sediment modeling using artificial neural networks, M.Tech Thesis [*Soil and Water Conservation Engineering & Irrigation and Drainage Engineering*] The G.B. Pant University of Agriculture and Technology, Pantnagar-263145, Uttarkhand, India.

Raju, M. M., Srivastava, R. K., Bisht, D. C. S., Sharma, H. C. and Kumar, A. (2011). Development of artificial neural-network-based models for the simulation of spring discharge. *Advances in Artificial Intelligence, Hindawi Publishing Corporation*, Volume 2011, Article ID:686258.

Refenes, A. N., and Vithlani, S. (1991). Constructive learning by specialization. Artificial Neural Networks, T. Kohonen et al., 923–929.

Sarkar, A., Raju, M.M., Kumar, A. (2010). Sediment runoff modeling using artificial neural networks. *Journal of Indian Water Resources Society, Vol. 30, No. 1* January, 2010.

Sato, A. (1991). An analytical study of the momentum term in a back-propagation algorithm. Artificial Neural Networks, T. Kohonen et al., 617–622.

Tayfur, G., and Guldal, V. (2005). Artificial neural networks for estimating daily total suspended sediment in natural streams. *Nordic Hydrology*, 37(1), 69–79.

Wu, T. H., Hall, J. A., and Bonta, J. V. (1993). Evaluation of runoff and erosion models. *Journal of Irrigation and Drainage Engineering, ASCE*, 119(4), 364–382.

Yu, P.S., Liu, C.L., and Lee T.Y. (1994). Application of a transfer function model to a storage runoff process. Stochastic and statistical methods in Hydrology and Environmental Engieeering. Vol. 3, 87–97.

Zurada. (1992). Introduction to artificial neural system, West publishing company. St. Paul, New York.

7

Ranking Forecasting Algorithms Using MCDM Methods: A Python Based Application

Swasti Arya, Mihika Chitranshi*, and Yograj Singh

Department of Mathematics, Lady Shri Ram College for Women, University of Delhi, India
E-mail: swasarya@gmail.com; mihichi1699@gmail.com; yograjchauhan26@gmail.com
*Corresponding Author

Abstract

To handle future uncertainty, it is critical to utilise time series forecasts while making decisions. Efficient forecasting is seen as a significant prerequisite for successful management and organization in a variety of social, knowledge, human and natural sciences and related fields of application. Various forecasting approaches have been proposed to deal with the increasing uncertainty and complexities associated with domain-specific forecasting problems. When deciding on a forecasting algorithm, decision-makers must consider various aspects of the prediction process, such as the duration of the forecast horizon, the aim of forecasting, the frequency, structure and nature of the data. The focus of this paper is to implement three MADM techniques namely, Analytic Hierarchy Process (AHP), Technique for Order of Preference by Similarity to Ideal Solution (TOPSIS) and VlseKriterijumska Optimizacija I Kompromisno Resenje (VIKOR) on a practical application using Python to propose an MCDM approach to evaluate and rank quantitative demand forecasting models. The rankings are based on error measurements like Mean Squared Error (MSE), Root Mean Squared Error (RMSE), etc. which play an instrumental role in forecasting algorithms. The conclusion outlines the best time-series method that complies with the given constraints.

Keywords: MCDM, AHP, TOPSIS, VIKOR, forecasting, time series, error measurement.

7.1 Introduction

Forecasts are an essential part of making logical decisions and constructing a game plan to tackle future uncertainties (Mehdiyev et al., 2016, p. 22–35). For efficient administration, several decisions need to be based on what the conditions will be in the near future. These could be social, economic, political, supply chain conditions to name a few. These conditions are extremely temperamental and may have significant fluctuations in a short time period. To illustrate, one might consider daily fluctuating stock prices or yearly sales figures in business. Or the annual temperature and the hourly wind speed in meteorology. Even in biomathematics, to record a heart's electrical activity at millisecond intervals (Cryer et al., 2008). To take into account the variability and deviations associated with the problem of forecasts, diverse forecasts methods have been proposed. In this context, researchers are confronted with the dilemma of choosing the best forecasting model, typically with the smallest error when compared with the real observation (Badulescu et al., 2018). To evaluate the forecasting methods several criteria can be taken into consideration. The structure and the nature of the available data, the time period for which the forecasts have been made, the frequency of the recorded observations, to name a few (Mehdiyev et al., 2016, p. 22–35). However, a single parameter or even error measurement falls short in evaluating various forecasting techniques which may just highlight a specific aspect or the role of outliers in a forecast analysis. This paper proposes the use of Multi-Criteria Decision Making (MCDM) techniques to evaluate different forecasting methods based on multiple criteria and ultimately choosing the best forecasting algorithm.

7.2 Review of Literature

7.2.1 Analytic Hierarchy Process (AHP)

The research work carried out by Thomas L. Satty (Satty, 1987, p. 161–176) resulted in the conception of the AHP technique. Although the method is analytical because of its name and the fact that it divides the abstract entity into its constituent elements, due to its ability to quantify and synthesize the multitude of factors in a hierarchy (Foreman et al., 2001), the method is essential (Russo et al., 2015). AHP has been widely used since its introduction, for example in the prioritization of evaluation used to determine the relative merit of a set of alternatives, quality management, especially in software, healthcare, resource allocation, benchmarking processes or a process system, machine selection (San and Tab canon, 1994), selection of industrial research and development projects and allocation of resources (Liberator, 1987). For the evaluation and justification of an advanced manufacturing system, AHP based on fuzzy numbers multi-attribute technique is also proposed (Duran et al., 2008, p. 1784–1797). The method has the distinct advantage of decomposing a decision problem into its constituent parts and constructing criteria hierarchies. The significance of each element (criterion) becomes clear here (Macharis et al., 2004, p. 307–317). Despite the popularity of the AHP, many authors have expressed concern about certain methodology issues and have observed some cases in which, when the AHP or some of its variants are used, ranking irregularities can occur.

7.2.2 Technique for Order of Preference by Similarity to Ideal Solution (TOPSIS)

TOPSIS is a unique method with an 'approach to identify an alternative in a multi-dimensional computing space that is closest to the ideal solution and farthest to the negative ideal solution' (Satty, 1987, p. 161–176). It has several benefits that make it stand out as a tool for decision-making. It has a simple, easy-to-use and programmable process. Regardless of the number of attributes (Anandan et al., 2017, p. 897–908), the number of steps remains the same. The method based on normalization techniques, distance measurements, dependence between alternatives, etc., has been revised in several forms. In his work (Chen, 2000, p. 1–9), Chen-Tung Chen suggested the vertex method to find the distance between PIS and NIS (Anandan et al., 2017, p. 897–908). Several AHP and TOPSIS, VIKOR and TOPSIS hybrid approaches and other methods have been identified in the literature to provide solutions to MCDM problems (Anandan et al., 2017, p. 897–908). A disadvantage is that the correlation of attributes is not considered by its use of Euclidean Distance. It is difficult, particularly with additional attributes, to weigh attributes and maintain consistency of judgment. TOPSIS finds its applications in fields such as supply chain management and logistics, design, engineering and manufacturing systems, business and marketing management, environmental management, human resources management and water resources management.

7.2.3 VIseKriterijumska Optimizacija I Kompromisno Resenje (VIKOR)

Opricovic (Opricovic, 1990) presented the VIKOR technique as a well-known MCDM approach to solve problems with contradictory parameters and provide a compromise solution. The method is implemented by ranking and selecting from a set of alternatives with conflicting criteria. A new model was proposed by Opricovic and Tzeng (Opricovic et al., 1990, p. 635–652), integrating VIKOR and TOPSIS for defuzzification within the multiple criteria decision-making model combining fuzzy and crisp criteria. Tzeng et al. (Opricovic et al., 2004, p. 514–529), used VIKOR, AHP and TOPSIS techniques to determine the best fuel alternatives in the technological development of buses. In addition, with TOPSIS, ELECTRE and PROMETHEE approach, Opricovic and Tzeng studied an extension of the VIKOR system and concluded that the VIKOR method is superior in meeting contradictory and non-commensurable attributes (Opricovic et al., 2007, p. 514–529). An extension was given to the definition of the VIKOR method to solve MADM problems with interval fuzzy numbers by (Sayadi et al., 2009, p. 2257–2262). In addition, it has been found that linguistic variables are the weights of attributes and ratings of alternatives (Chen, 2000, p. 1–9). In several fields, such as architecture, mechanical and transportation engineering and manufacturing, the method has a wide range of applications.

7.2.4 Time Series Analysis

Forecasting techniques for a singlevariable are meant to look for patterns in time series and thus, make predictions based on these values. The primary form of forecasting a time series is based on weighted averages of earlier values taking into consideration seasonal cycles and time trends (Zhu et al., 2019, p. 65–82). The Russian statistician and economist, E. Slutsky (1927), and the British statistician G.U. Yule (1921, 1926, 1927) simultaneously suggested characterising time series by using autoregressive (AR) or moving-average (MA) processes or combined autoregressive moving-average (ARMA) processes (Nerlove et al., 1990, p. 22–35). Udny Yule's seminal paper was the first application of ARMA analysis to real-life data which illustrated a method to do away with the assumption of periodicity (Nielsen, 2019). George Box, a pioneering statistician developed the Box-Jenkins method (1970) in a statistics textbook Time Series Analysis: Forecasting and Control (Wiley). The method and the book, now in its fifth edition remain popular. This later acted as the base for the auto-regressive integrated moving average (ARIMA) method (Box et al. 1994; Sheu 2010). Tratar et al. (2016) developed a smoothing and forecasting method to effectively handle both additive and multiplicative seasonality. Since then, several variations have been conceptualised and applied to practical fields. The idea of combining forecasts instead of choosing the best one originated in 1969, which was looked down upon by traditional statisticians (Elmunim et al., 2015). Even though medicine had a slower pace to integrate the concepts of time series, it led to the origination of the discipline of demography and life tables (John Graunt, 1662). It has proved useful in medical tests like electrocardiograms. By the late 19th century, the fields of weather forecasting and economic growth garnered attention and revealed cyclical patterns in the weather and the stock market. Mechanical trading was just an application of time series forecasting (Richard Dennis, 1970s and 1980s) which then developed into the use of Artificial Intelligence (1980s) for forecasting. With the boost of data quantity, methods like neural networks and fuzzy adaptations are being worked upon. Wu (2012) proposed a forecasting method combining time series analysis and neural networks to forecast dynamic relief needs in earthquake-affected areas. (Zhu et al., 2019, p. 65–82).

7.3 Error Measurements and Forecasting

Forecasting is a method that takes historical data as input to make logical estimates that are instrumental in predicting the direction of future trends. Regular patterns in time series are extrapolated to determine forecasts using quantitative methods that approach the problem with a mathematical aspect (Badulescu et al., 2018). The crucial parameters to judge the accuracy of the forecasted values are error measurements. The objective of this paper is to use these error measurements as the criteria in an MCDM problem, to filter out the best forecasting algorithm. Different forecasting methods result in different types of statistical error measurements, which are described in Table 7.1 (Badulescu et al., 2018).

In this paper we will only be using Mean Error, Mean Absolute Error, Root Mean Squared Error, Mean Percentage Error, Root Mean Squared Percentage Error, Mean Absolute Percentage Error and R^2. These will act as the parameters on which we evaluate the choices or alternatives we have. The alternatives are five different time series forecasting methods. Three of which are traditional methods whereas two are hybrid methods. A brief description of the forecasting methods is given below.

Table 7.1 Descriptions of error measurements

Name of the Error	Formula	Description
Mean Error (ME)	$\frac{1}{n}\Sigma(y_i - \hat{y})$	The simplest form of error measurement. It indicates if the forecasted values are biased. Oftentimes, a positive error counterbalances a negative error at different data points and results in inaccuracies.
Mean Absolute Error (MAE)	$\frac{1}{n}\Sigma|y_i - \hat{y}|$	This is an improvement of Mean Error as here the absolute difference prevents the counterbalancing of positive and negative error values. However, when there are large outliers, these error measurements may give faulty results as the mean is skewed.
Mean Percentage Error (MPE)	$\frac{100\%}{n}\Sigma\left(\frac{y_i - \hat{y}}{y_i}\right)$	Since this error measurement method is based on the actual values and not the absolute value, it gives a better outcome in terms of the relative size and the direction of the bias.
Root Mean Squared Percentage Error (RMSPE)	$\sqrt{\frac{1}{n}\Sigma\left(\frac{y_i - \hat{y}}{y_i}\right)^2}$	Owing to the squaring of the value, it results only in positive values. Hence, it only provides a relative size of the error present. This error measurement method is not capable of handling large outliers.
Root Mean Squared Error (RMSE)	$\sqrt{\frac{1}{n}\Sigma(y_i - \hat{y})^2}$	Owing to the squaring function, large errors are given extra weightage. However, it is a good representation of the size of the error.
Mean Absolute Percentage Error (MAPE)	$\frac{100\%}{n}\Sigma\left|\frac{y_i - \hat{y}}{y_i}\right|$	It is the most widely used error measurement method while forecasting demand. As it is expressed as a percentage, it is free from any unit. The lower boundary for MAPE is 0, however, it does not have an upper boundary. It increases when there is an asymmetrical distribution and in the presence of outliers.
R-squared (R^2)	$1 - \frac{\Sigma(y_i - \hat{y}_i)^2}{\Sigma(y_i - \overline{y})^2}$	Also known as the Coefficient of Determination, it tells how good will the future values, predicted by the algorithm, be. Typically, closer the value of R-Squared is to 1, the better is the forecasted value. Nonetheless, there may be certain instances where even a negative value of R^2 is achieved.

7.3.1 Holt-Winter

Is a statistical short-term forecasting method that provides results with the same accuracy over time (Elmunim et al., 2015). Holt Winter method uses exponential smoothening to model the average, the trend and the seasonality of a time series to forecast typical values for the present and the future. The Holt-Winter method has two models namely, the additive model and the multiplicative model. The additive model is represented as Equation (7.1) (Razali et al., 2018):

$$\hat{y}_{t+h|t} = l_t + hbt + s_{t+h-m(k+1)}$$
$$l_t = \alpha(y_t - s_{t-m}) + (1 - \alpha)(l_{t-1} + b_{t-1})$$
$$b_t = \beta^*(l_t - l_{t-1}) + (1 - \beta^*)b_{t-1}$$
$$s_t - \gamma(y_t - l_{t-1} - b_{t-1}) + (1 - \gamma)s_{t-m} \qquad (7.1)$$

And the multiplicative model is represented as given by Equation (7.2):

$$\hat{y}_{t+h|t} = (l_t + hbt) + s_{t+h-m(k+1)}$$
$$l_t = \alpha \frac{y_t}{s_{t-m}} + (1 - \alpha)(l_{t-1} + b_{t-1})$$
$$b_t = \beta^*(l_t - l_{t-1}) + (1 - \beta^*)b_{t-1}$$
$$s_t - \gamma \frac{y_t}{(l_{t-1} - b_{t-1})} + (1 - \gamma)s_{t-m} \qquad (7.2)$$

The Holt-Winters seasonal method comprises the forecast equation and three smoothing equations – one for the level l_t, one for the trend b_t, and one for the seasonal component s_t, with corresponding smoothing parameters α, β^* and γ. We use m to denote the frequency of the seasonality, i.e., the number of seasons in a year.

7.3.2 Autoregressive Integrated Moving Average (ARIMA)

Is one of the most widely used forecasting algorithms. It is a generalization of ARMA (Xu et al., 2010, p. 4313–4317) where ARMA is a combination of AR and MA. ARIMA models are used to outline the autocorrelations in the data present. Some special cases of ARIMA models are Random-walk, Random-Trend models, Autoregressive models and Exponential Smoothening models. The major drawback of ARIMA is that it assumes a linear form of the model and no non-linear patterns are captured by it. Oftentimes, real-world complex problems cannot be modeled accurately through linear model approximations

(Zhang, 2003, p. 159–175). The time series has the form given by Equation (7.3):

$$y_t = \theta_0 + \phi_1 y_{t-1} + \phi_2 y_{t-2} + \cdots + \phi_p y_{t-p} + \varepsilon_t - \theta_1 \varepsilon_{t-1} - \theta_2 \varepsilon_{t-2} - \cdots - \theta_q \varepsilon_{t-q} \quad (7.3)$$

where y_t denotes the actual values whereas ε_t denotes the random error in time t. The model parameters are represented by ϕ_i $(i = 1, 2, \cdots, p)$ and θ_i $(j = 1, 2, \cdots, q)$. Here the random errors, ε_t, are assumed to be identically distributed with a mean of zero and a constant variance of σ^2. The random errors are also assumed to be independent.

7.3.3 SARIMA

Is the most popular method for forecasting seasonal time series. It is an extension of the ARIMA model. It has an additional parameter for including the period of seasonality in the model (Alencar et al., 2018). The SARIMA model is described mathematically by Equation (7.4):

$$\emptyset_p(B) \Phi_p\left(B^S\right) \nabla^d \nabla_S^D y_t = \emptyset_q(B) \Theta_Q\left(B^S\right) \varepsilon_t \quad (7.4)$$

where ϕ is the regular AR polynomial of order p, Φ is the seasonal AR polynomial of order P, θ is the regular MA polynomial of order q, Θ is the seasonal MA polynomial of order Q, ∇^d is the differentiating operator, ∇_S^D is the seasonal differentiating operator, y_t is the wind speed at a time t, ε_t is the residual error at time t and B is the backshift operator as $B^k(y_t) = y_{t-1}$.

7.3.4 ARIMA integrating Single Judgement Adjustment

Is a hybrid forecasting method that includes human judgement. The experience and knowledge of an expert are combined in a systemised format in addition to the mathematical or statistical predictions, which are made using traditional algorithms like ARIMA. These prove to be good fits to the models but may occasionally be inaccurate if the human expert were to make any mistakes in the mathematical forecast.

7.3.5 ARIMA integrating Collaborative Judgement Adjustment

Is a hybrid forecasting method that includes a team comprising three experts. A similar process is followed as that in ARIMA integrating Single Judgement Adjustment, the only difference being in the number of human experts taken into consideration. This might also reduce the drastic effect of a human error made by a single expert. There are several methods to integrate the observations and decisions of multiple people.

7.4 Multi Criteria Decision Making

Even though decision-making is an extremely intuitive process, a Decision Maker (DM) usually follows a set collection of steps to arrive at the answer. The first step is to recognise the need to make a decision. Following which certain parameters known as criteria are set, against which the available alternatives are evaluated. To outline the algorithm, the process of making a decision has four main steps – formation of the problem statement, setting the criteria, evaluating the alternatives available and determination of the ideal alternative. Be it purchasing a new automobile or deciding the location of a new factory, it all boils down to evaluating different options and ultimately selecting the best one. Oftentimes one might desire an alternative that satisfies conflicting criteria.

Consider purchasing an automobile, the better the fuel efficiency, higher the price. A DM would still like to attach different importance to the two criteria of fuel efficiency and price, in order to find an alternative that makes the correct trade-off to secure an automobile with low price and high fuel efficiency. Each criterion could have a different unit of measurement. In the example stated above, the fuel efficiency of an automobile is measured in liters per 100 kilometres whereas cost is measured in currency. Clearly, the DM will evaluate his choices on multiple criteria and hence the name, MCDM.

MCDM can be considered as a natural outgrowth of the application side of Operational Research. MCDM and traditional Operational Research techniques usually focus on maximizing or minimizing a utility function in the presence of constraints (Zhu et al., 2019, p. 65–82). To facilitate a systemised approach to decision making, Hwang and Yoon (1981) suggested that MCDM problems can be classified into two main categories:

7.4.1 Multiple Attribute Decision Making (MADM)

A branch of MCDM is utilised when the number of alternatives is fixed and defined before beginning the process. The options are limited in number. MADM requires both inter- and intra-attribute comparisons. MADM refers to making preference decisions (e.g. evaluation, prioritization, selection) (Macharis et al., 2004). It is further possible to classify MADM problems into two sub-branches namely Multiple Attribute Utility Theory (MAUT) and outranking techniques (Macharis et al., 2004). MAUT approaches include deciding the desires of the Decision Maker, represented by using an effective utility function as a hierarchical structure. A limitation of MAUT techniques, that has been highlighted is that it ignores the interaction among the different criteria and its subsequent impact (Duran et al., 2008, p. 1784–1797). For example, the decision behind purchasing a smartphone that satisfies certain criteria is a MADM problem, wherein there are a finite number of alternatives to choose from.

7.4.2 Multiple Objective Decision Making (MODM)

The other branch of MCDM problems is used when there is not a specified number of alternatives before beginning the process. Using multiple objective programming a set of alternatives is defined. The possible number of alternatives in this case can be many. In other words, MODM problems prove instrumental when the best alternative needs to be constructed instead of selected (Duran et al., 2008, p. 1784–1797). Oftentimes, this results in a considerable increase in the number of dimensions beyond the handling capacity, which in turn increases multiple computational costs. This leads to financial constraints. For example, automobile manufacturers wish to design a car that minimizes production cost and maximizes riding comfort and fuel economy. After the process, several alternatives are created that fit the conflicting criteria.

7.5 MCDM Methods

Among the various MODM and MADM methods, this paper focuses on three MADM methods namely, Analytic Hierarchy Process (AHP), Technique for Order of Preference by Similarity to Ideal Solution (TOPSIS) and VlseKriterijumska Optimizacija I Kompromisno Resenje (VIKOR).

7.5.1 The AHP Method

AHP is a method to solve MADM problems, specifically to determine the weights of the criteria. Saaty (1977,1980) was the first to propose AHP (Anandan et al., 2017, p. 897–908). A unique factor about AHP is that it

Table 7.2 Ratio scale for AHP

Intensity	1	3	5	7	9	2,4,6,8
Linguistic	Equal	Moderate	Strong	Demonstrated	Extreme	Intermediate Values

pays importance to the human experience and knowledge as well. The AHP consists of four broad parts- decomposing the problem to set up a hierarchical structure with interrelated elements, forming the reciprocal matrix by pair-wise comparison, estimating the relative weights after synthesising priorities and eventually determining the best alternative by taking into account the aggregate relative weights. In this paper, we make use of AHP to determine the criteria weights only.

Steps of Methods:

1. Set up a hierarchical structure by analysing the problem statement in criteria, sub-criteria and alternatives. The objective is level 1. The next level has the criteria, each of which is again divided into sub-criteria.
2. Compare all elements pairwise with respect to the objective. Each criterion is compared to every other criterion on just a single property without concern for other properties or other elements. Each comparison is carried out on the Ratio Scale as given in Table 7.2.

Assuming there are n criteria or attributes that need to be taken into consideration. Suppose Attribute 1 is extremely important than Attribute 2, then Attribute is said to be nine times more important than Attribute 2. Consequently, Attribute 2 is 1/9 times more important than Attribute 1. When Attribute 1 is compared with itself then intensity 1 is assigned. Similarly, a matrix X_{ij} is composed of these values from the ratio scale. This is called the pair-wise matrix.

3. Calculate the normalised principal eigenvector.
- Calculate the normalised pairwise matrix

$$X_{ij}^* = \frac{x_{ij}}{\sum_{i=1}^{n} x_{ij}} \; for \; j = 1, 2, \ldots n.$$

- Calculate criteria weights by averaging all the elements in the row

$$w_i = \frac{1}{n} * \left(\sum_{j=1}^{n} x_{ij} \right).$$

Table 7.3 Random index or consistency index for AHP

n	1	2	3	4	5	6	7	8	9	10
R.I.	0.00	0.00	0.58	0.90	1.12	1.24	1.32	1.41	1.45	1.49

- In order to check whether the calculated weights are correct we calculate Consistency Index. Take the original pairwise matrix X_{ij} and multiply with the criteria weight to each column.

$$X_{ij}^c = w_i * X_{ij}$$

- Calculate weighted sum value

$$S_i = \sum_{j=1}^{n} X_{ij}^c$$

- Calculate $\lambda_{max} = \frac{1}{n} * \sum_{i=1}^{n} \left(\frac{S_i}{w_i}\right)$
- Calculate Consistency Index $C.I. = \frac{\lambda_{max} - n}{n-1}$ where n is the number of criteria
- Calculate Consistency Ratio

$$C.R. = \frac{C.I.}{R.I.}$$

where $R.I.$ is Random Index or the Consistency Index of randomly generated pairwise matrix from Table 7.3.

- If $C.R. < 0.10$ then we may assume that the criteria weights are sufficiently consistent and proceed with the process.

7.5.2 The TOPSIS Method

TOPSIS is an approach to determine the best alternative in a multi-attribute scenario which was first put forth by Hwang and Yoon in 1981. This is achieved by evaluating the distance of the alternative to the ideal solutions. The one with the shortest distance from the positive ideal solution and the longest distance from the negative ideal solution is ranked the best option. This is done using a Relative Closeness Coefficient or Similarity Index. The alternative which has the highest Similarity Index is judged to be the best. Although any distance measuring family can be used to find the distance from the Positive Ideal Solution or the Negative Ideal Solution, Euclidean Distance remains the most popular one. Some examples of other distance measures are Soergel, Canberra, Gower, City Block, Minkowski, Lorentzian, etc. This paper uses Euclidean distance to calculate the separation of the alternative to the ideal solutions.

Steps of method:

1. TOPSIS assumes that there are m alternatives, n attributes/criteria and the score of each option with respect to each criterion is known.
2. Let J be the set of beneficial criteria (more is better) and J' be the set of non-beneficial criteria (less is better)
3. Let x_{ij} be the score of option i with respect to criterion j. We have a matrix $X = (x_{ij}) = m \times n$ matrix.
4. Construct the normalized decision matrix using the formula: $r_{ij} = \frac{x_{ij}}{\sqrt{\sum x_{ij}^2}}$ for $i = 1, 2, 3, \ldots, m$; $j = 1, 2, 3, \ldots, n$
5. Given that the set of weights for each criterion w_j for $j = 1, 2, 3, \ldots, n$. Multiply each column of the normalized decision matrix by its associated weight w_j. This generates the weighted normalised decision matrix. Each element of which is represented as

$$v_{ij} = w_{ij} * r_{ij}$$

6. Calculate the ideal best and the ideal worst value for each criterion. For a beneficial criterion, the best value is the maximum value in the criteria whereas the worst value is the minimum value in the criteria. For a non-beneficial criterion, the best value is the minimum value in the criteria whereas the worst value is the maximum value in the criteria. For a simpler understanding, refer to Table 7.4.
7. For calculating separation measure from each alternative, we use Euclidean distance, given by the formula: Euclidean distance $d(a_i, b_i)$ between points a_i and $b_i = \sqrt{(\sum_{i=1}^{p} (a_i - b_i)^2)}$
8. The distance from the ideal best solution $S_i^+ = \frac{\sum_j |v_{ij} -- V_j^+|}{m}$ for $i = 1, 2, 3, \ldots, m$ The distance from the ideal worst solution S_i^- = $\frac{\sum_j |v_{ij} -- V_j^-|}{m}$ for $i = 1, 2, 3, \ldots, m$

Table 7.4 Best and worst values for criteria for TOPSIS

	Beneficial Criteria	Non-Beneficial Criteria
Best Value (V_j^+)	$(v_{ij})_{max}$	$(v_{ij})_{min}$
Worst Value (V_j^-)	$(v_{ij})_{min}$	$(v_{ij})_{max}$

9. Calculate the relative closeness P_i by the formula:

$$P_i = \frac{S_i^-}{S_i^- + S_i^+} \quad 0 < P_i < 1$$

10. The alternatives are ranked according to their relative scores. The higher the value of P_i, better the rank is. In other words, the one with P_i value closest to 1 is the best alternative.

7.5.3 The VIKOR Method

VIKOR, a Serbian phrase, translates to Multi-Criteria Optimization and Compromise Solution. L_p- metric which is used in the compromise programming method laid the foundation for the development of the multiple attribute merit for compromise ranking or VIKOR. The primary groundwork for VIKOR was done by Yu 1973 and Zeleny 1982. Opricovic 1998 and Tzeng later popularised the method (Chen, 2000, p. 1–9). VIKOR is an exceptional decision-making method to tackle situations where the Decision Maker/s are unable to express or incorporate their preferences prior to the start of the system design. After the application of the method, the compromise solution obtained can be used as a basis for negotiations. The VIKOR method was started with the form of L_p- metric given in Equation (7.5).

$$L_{p,k} = \left\{ \sum_{j=1}^{n} [w_j \left(f_j^* - f_{k_j}\right)/\left(f_j^* - f_i^-\right)]^p \right\}^{\frac{1}{p}}, \quad 1 \le p \le \infty; \quad (7.5)$$

$$k = 1, 2, \ldots n \quad (7.6)$$

Steps of Methods:

1. Set up an $m \times n$ decision matrix namely X_{ij}, where m corresponds to the number of alternatives and n corresponds to the number of criteria and i = 1, 2, ... m and j = 1, 2, ... n. The weightage of each criteria is a priori and represented by W_j.
2. Identify beneficial and non-beneficial criteria. A beneficial criterion is one whose higher value is desired whereas a non-beneficial criterion is one whose lower value is desired. A non-beneficial criterion is also called a cost criterion.
3. Calculate the best and worst values for each criterion. For a beneficial criterion, the best value is the maximum value in the criteria whereas the

Table 7.5 Best and worst values for criteria for VIKOR

	Beneficial Criteria	Non-Beneficial Criteria
Best Value (x_i^+)	$(x_{ij})_{max}$	$(x_{ij})_{min}$
Worst Value (x_i^-)	$(x_{ij})_{min}$	$(x_{ij})_{max}$

Table 7.6 Best and worst values of S_i and R_i

Best Value	$S^* = \min_i S_i$	$R^* = \min_i R_i$
Worst Value	$S^- = \max_i S_i$	$R^- = \max_i R_i$

worst value is the minimum value in the criteria. For a non-beneficial criterion, the best value is the minimum value in the criteria whereas the worst value is the maximum value in the criteria. For a simpler understanding, refer to Table 7.5.

4. Calculate Unity Measure S_i using the below mathematical formula

$$S_i = \sum_{j=1}^{m} \left[W_j * \frac{x_i^+ - x_{ij}}{x_i^+ - x_i^-} \right]$$

5. Calculate Individual Regret (R_i):

$$R_i = \max_j \left[W_j * \frac{x_i^+ - x_{ij}}{x_i^+ - x_i^-} \right]$$

6. Calculate the best and the worst values of S_i and R_i as given in Table 7.6.

7. Calculate the value of Q_i

$$Q_i = \nu * \frac{S_i - S^*}{S^- - S^*} + (1 - \nu) * \frac{R_i - R^*}{R^- - R^*}$$

where ν is the weight for the strategy of maximum group utility. Normally taken as $\nu = 0.5$.

8. Rank the alternatives based on the values of Q_i. The alternative with the minimum Q_i value is given rank 1 and the alternative with the maximum Q_i value is given the last rank. The Decision Maker can make his/her decision based on these ranks.

9. Using the VIKOR method, a compromise solution is proposed based on two conditions called C1 and C2:

- **C1: Acceptable Advantage** We check if this inequality is satisfied: $Q(A^2) - Q(A^1) \geq DQ$ where $Q(A^i)$ is the Q value of the alternative with rank i and $= \frac{1}{m-1}$, where m is the number of alternatives.
- **C2: Acceptable Stability in Decision Making** Alternative with rank 1 according to Q_i must also be the alternative that is ranked 1 when ranked by the value of S and/or R.

If any one of the conditions is not satisfied, then a set of compromise solutions is proposed, which consists of:

1. Alternatives with rank 1 and 2 if only condition C2 is not satisfied, or
2. Alternatives with rank 1, 2, ... M if condition C1 is not satisfied; and M is determined by the relation $Q(A^M) - Q(A^1) < DQ$ for maximum M (the positions of these alternatives are 'in closeness').

This compromise solution is stable within a decision-making process, which could be: 'voting by majority rule' (when $\nu > 0.5$ is needed) or 'by consensus' ($\nu \approx 0.5$) or 'with vote' ($\nu < 0.5$). For an alternative to having added advantage over another, the minimum difference between their Q values must be DQ.

7.6 Framework of the Problem

The objective of the problem statement is to analyse and rank quantitative demand forecasting models. The rankings are based on error measurements which serve as criteria and the forecasting methods serve as alternatives or attributes. The error measurements are specified with their notations in Table 7.7.

Only R-squared error has been taken as the beneficial criteria and the rest as non-beneficial criteria. In other words, a higher value of R-squared error is desired whereas a lower value for the other criteria.

The inspiration to use MCDM methods to analyse the aforementioned topic was derived from (Badulescu et al., 2018). This paper aims at recreating the same problem statement and data points, examining it through AHP, TOPSIS and VIKOR.

For the purpose of this study the alternatives along with their representation, which are used in this paper henceforth, are elaborated in Table 7.8.

7.7 Implementation Using Python Programming Language

Table 7.7 Description of criteria

Criteria	Explanation of the Criteria
C_1	Mean error (ME)
C_2	Mean absolute error (MAE)
C_3	Mean percentage error (MPE)
C_4	Root mean squared percentage error (RMSPE)
C_5	Root mean squared error (RMSE)
C_6	Mean absolute percentage error (MAPE)
C_7	R-squared (R^2)

Table 7.8 Description of the alternatives

Representation	Explanation of the Alternatives
SC_1	Holt-Winter
SC_2	ARIMA
SC_3	SARIMA
SC_4	ARIMA integrating Single Judgement Adjustment
SC_5	ARIMA integrating Collaborative Judgement Adjustment

7.7 Implementation Using Python Programming Language

The multi criteria decision matrix is represented as illustrated in Table 7.9.

7.7.1 Determining the criteria weights using AHP

Using the Ratio Scale in Table 7.2, the following pairwise comparison matrix was created for the criteria, using the representations as explained earlier. The pairwise comparison matrix can be seen in Table 7.10.

Table 7.9 Decision matrix

Weights	0.049752	0.099672	0.164711	0.082165	0.170663	0.301525	0.131512
Alternatives	C_1	C_2	C_3	C_4	C_5	C_6	C_7
SC_1	36.94	75.99	92.53	3.27	7.35	5.78	0.61
SC_2	1.33	112.17	129.7	0.51	9.74	8.19	0.24
SC_3	17.5	92	107.11	1.7	7.97	6.68	0.48
SC_4	−52.42	97.42	171.94	−3.29	11.06	6.66	−0.34
SC_5	5.58	33.25	42.85	0.47	3.04	2.36	0.92

Table 7.10 Pairwise comparison matrix

Criteria	C_1	C_2	C_3	C_4	C_5	C_6	C_7
C_1	1.000	0.200	0.250	0.333	0.200	0.200	1.000
C_2	5.000	1.000	0.333	1.000	0.333	0.333	1.000
C_3	4.000	3.000	1.000	3.000	1.000	0.333	1.000
C_4	3.000	1.000	0.333	1.000	0.333	0.200	1.000
C_5	5.000	3.000	1.000	3.000	1.000	0.333	1.000
C_6	5.000	3.000	3.000	5.000	3.000	1.000	1.000
C_7	1.000	1.000	1.000	1.000	1.000	1.000	1.000

Table 7.11 Criteria weights calculated using AHP

Criteria	Weight
C_1	0.049752
C_2	0.099672
C_3	0.164711
C_4	0.082165
C_5	0.170663
C_6	0.301525
C_7	0.131512

Following which, the pairwise comparison matrix was used as input to a Python code which resulted in the calculation of the following criteria weights and their notation is elaborated in Table 7.11.

The Consistency Index was determined as 0.12652910114 and the Consistency Ratio was calculated as 0.09585537965596594, which is less than 0.1, implying that the calculated criteria weights were consistent. The calculated criteria weights were then used to rank the alternatives.

7.7.2 Ranking Alternatives using TOPSIS method

The weighted normalized decision matrix which is calculated according to the aforementioned steps is specified in Table 7.12.

The set of Positive Ideal Solution (PIS) and Negative Ideal Solution (NIS) hence obtained are tabulated in Table 7.13.

The Relative Closeness P_i is calculated and alternatives are ranked according to the rule that the one with a higher value of P_i gets a better rank. Alternatives sorted according to their rank are given in Table 7.14.

7.7 Implementation Using Python Programming Language

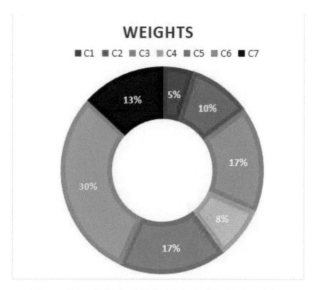

Figure 7.1 Criteria weights as calculated using AHP.

SC_5 (ARIMA Integrating Collaborative Judgement Adjustment) turns out to be the best alternative here. Followed by SC_1 (Holt-Winter). The least favourable alternative is SC_2 (ARIMA).

Table 7.12 Weighted normalised decision matrix

Alterna	C_1	C_2	C_3	C_4	C_5	C_6	C_7
SC_1	0.027545	0.039156	0.058336	0.053857	0.067634	0.124795	0.062988
SC_2	0.000992	0.057798	0.08177	0.0084	0.089626	0.176829	0.024782
SC_3	0.013049	0.047405	0.067528	0.027999	0.073339	0.144227	0.049564
SC_4	−0.03909	0.050198	0.1084	−0.05419	0.101772	0.143795	−0.03511
SC_5	0.004161	0.017133	0.027015	0.007741	0.027974	0.050954	0.094998

Table 7.13 Positive ideal solution and negative ideal solution

	C_1	C_2	C_3	C_4	C_5	C_6	C_7
Positive Ideal	0.027545	0.057798	0.1084	0.053857	0.101772	0.176829	0.094998
Negative Ideal	−0.03909	0.017133	0.027015	−0.05419	0.027974	0.050954	−0.03511

118 *Ranking Forecasting Algorithms Using MCDM Methods*

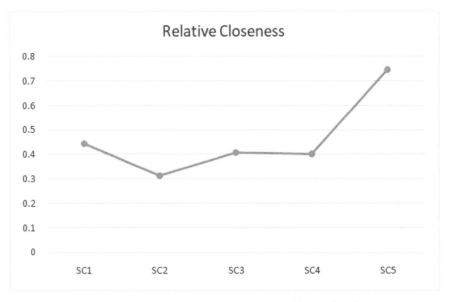

Figure 7.2 Line graph of relative closeness coefficients in TOPSIS.

7.7.3 Ranking Alternatives using VIKOR method

Values of S_i, R_i and Q_i are evaluated according to the formulae specified earlier are noted in Table 7.15.

Ranks based on S_i, R_i and Q_i are given in Table 7.16.

Since both the conditions, acceptable advantage (C1) and acceptable stability in decision making (C2), are satisfied, the final ranking of alternatives is done based on the value of Q_i.

Thus, SC_5 (ARIMA integrating collaborative judgement adjustment) is the best alternative. Followed by SC_1 (Holt-Winter). and SC_2 (ARIMA) is the least ranked alternative.

Table 7.14 Ranked alternatives based on TOPSIS

Alternatives	Relative Closeness	Rank
SC_5	0.745911	1
SC_1	0.444045	2
SC_3	0.406907	3
SC_4	0.400403	4
SC_2	0.313023	5

Table 7.15 Values of S_i, R_i and Q_i for VIKOR

Alternatives	C_1	C_2	C_3	C_4	C_5	C_6	C_7	Si	Ri	Qi
SC_1	36.94	75.99	92.53	3.27	7.35	5.78	0.61	0.550237	0.176881	0.597631
SC_2	1.33	112.17	129.7	0.51	9.74	8.19	0.24	0.803082	0.301525	1
SC_3	17.5	92	107.11	1.7	7.97	6.68	0.48	0.631881	0.223428	0.739937
SC_4	−52.42	97.42	171.94	−3.29	11.06	6.66	−0.34	0.770324	0.222394	0.824098
SC_5	5.58	33.25	42.85	0.47	3.04	2.36	0.92	0.079387	0.047095	0.049426

Table 7.16 Ranks based on S_i, R_i and Q_i for VIKOR

Alternatives	Rank Based on S_i	Rank Based on R_i	Rank Based on Q_i
SC_1	2	2	2
SC_2	5	5	5
SC_3	3	4	3
SC_4	4	3	4
SC_5	1	1	1

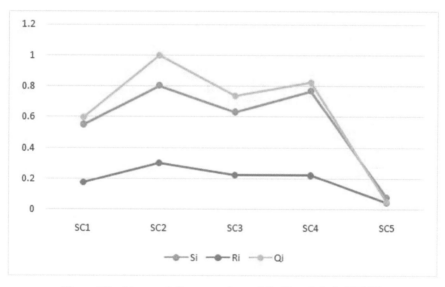

Figure 7.3 Line graph for comparison of S_i, R_i and Q_i in VIKOR.

7.8 Result Analysis

Both the ranking methods, TOPSIS and VIKOR have generated the same rankings for all five attributes. The result highlights that SC_5, that is, ARIMA

Integrating Collaborative Judgement Adjustment is the most suitable forecasting algorithm given the seven error measurements. The second most impactful parameter is SC_1 or Holt-Winter. Followed by, SARIMA model or SC_3. Next in rankings are SC_4 or ARIMA Integrating Single Judgement Adjustment and SC_2 or ARIMA model. The two methods generate the same results in terms of the ranks. The criteria weights taken for this implementation greatly depended on the perception of the DM. Any value changes in the pair-wise comparison matrix for AHP would have resulted in a different set of criteria weights, resulting in a different ranking for the given alternatives. Thus, the values for the pairwise comparison matrix for AHP in this paper were fixed after careful consideration.

7.9 Conclusion

This paper proposes the use of three MCDM approaches namely AHP, TOPSIS and VIKOR to aid in the selection of the most suitable demand forecasting system. Typical quantitative regression and hybrid demand forecasting models provide alternative forecasting techniques. The goodness-of-fit of the forecasts, calculated as the error measurements between forecasts and actual sales, is the criterion by which they are judged. The weights of the criteria are evaluated using the AHP method upon which the rankings are strongly dependent.

This approach has been implemented using lightweight codes and with as little complexity and runtime as possible in the Python programming language. For a better visual understanding, the weightage of the criteria obtained using AHP, the Relative Closeness in TOPSIS and the values of S_i, R_i and Q_i in VIKOR method have been analysed using suitable charts and graphs given in Figures 7.1, 7.2 and 7.3. The results of this study outline that ARIMA Integrating Collaborative Judgement Adjustment is the most favorable forecasting method in the given scenario. The results gained in both the approaches indicate close association thus justifying that the proposed approach can be applied to similar datasets of MCDM problems in similar scenarios.

The future scope of the research will explore the dimensions of the nature of forecasting algorithms being as highly qualitative or quantitative. A study with a larger data set in terms of the number of forecasting methods can be conducted. For future studies, the inter-relation among the different error measurements should also be taken into consideration and subsequently, the use of other multi-criteria decision-making methods should also be employed.

The impact of human expertise and human adjustment to the forecasting methods seems like an interesting area of further research. Moreover, the significance of the human experience could also be tested and experimented with.

7.10 Appendix

Figures 7.4, 7.5 and 7.6 are snippets of the codes programmed in Python given for reference. The codes have been kept simple and logical so that even non-coders can understand them without much difficulty. The variable names for the data frames and operations are self-explanatory. Certain outputs, where necessary have also been included in the figures.

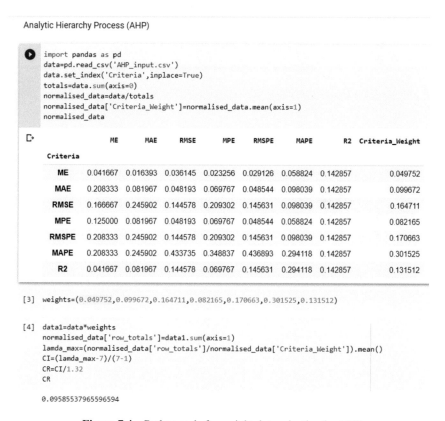

Figure 7.4 Python code for weight determination by AHP.

Technique for Order of Preference by Similarity to Ideal Solution (TOPSIS)

```python
import numpy as np
import pandas as pd
df=pd.read_csv('BookChapter_Input.csv')
df.set_index('Error Measurements',inplace=True)
weights=(0.049752,0.099672,0.164711,0.082165,0.170663,0.301525,0.131512)
df1=np.square(df)
totals=np.sqrt(df1.sum(axis=0))

normailised_df=df/totals
weighted_normalised_df=normailised_df*weights
wt_nrm_df=weighted_normalised_df.copy(deep=True)
weighted_normalised_df.loc['Max']=weighted_normalised_df.max(axis=0)
weighted_normalised_df.loc['Min']=weighted_normalised_df.min(axis=0)
weighted_normalised_df.loc['Beneficial']=(0,0,0,0,0,0,1)
weighted_normalised_df.loc['Non-Beneficial']=(1,1,1,1,1,1,0)
weighted_normalised_df.loc['Positive Ideal']=(weighted_normalised_df.loc['Beneficial']*weighted_normalised_df.loc['Max'])
                                            +(weighted_normalised_df.loc['Non-Beneficial']*weighted_normalised_df.loc['Min'])

weighted_normalised_df.loc['Negative Ideal']=(weighted_normalised_df.loc['Beneficial']*weighted_normalised_df.loc['Min'])
                                            +(weighted_normalised_df.loc['Non-Beneficial']*weighted_normalised_df.loc['Max'])

positive_ideal_df=(np.square(wt_nrm_df-weighted_normalised_df.loc['Positive Ideal']))
negative_ideal_df=(np.square(wt_nrm_df-weighted_normalised_df.loc['Negative Ideal']))
positive_ideal_df['Separation Ideal Solution']=np.sqrt(positive_ideal_df.sum(axis=1))
negative_ideal_df['Separation Ideal Worst']=np.sqrt(negative_ideal_df.sum(axis=1))
negative_ideal_df['Relative Closeness']=negative_ideal_df['Separation Ideal Worst']/(negative_ideal_df['Separation Ideal Worst']
                                            +positive_ideal_df['Separation Ideal Solution'])

negative_ideal_df['Relative Closeness']

Error Measurements
Holt-Winter            0.444045
ARIMA                  0.313023
SARIMA                 0.406907
Single Adjustment+ARIMA 0.400403
Collab Adjustment+ARIMA 0.745911
```

Figure 7.5   Python code for TOPSIS.

VIseKriterijumska Optimizacija I Kompromisno Resenje (VIKOR)

```python
import pandas as pd
import numpy as np
df=pd.read_csv('BookChapter_Input.csv')
df.set_index('Error Measurements',inplace=True)
df_new = df.copy()
df_extended = df.copy()
df_new.loc['Max']=df_new.max(axis=0)
df_new.loc['Min']=df_new.min(axis=0)
df_new.loc['Beneficial']=(0,0,0,0,0,0,1)
df_new.loc['Non-Beneficial']=(1,1,1,1,1,1,0)
df_new.loc['best_Xi']=(df_new.loc['Beneficial']*df_new.loc['Max'])
                      +(df_new.loc['Non-Beneficial']*df_new.loc['Min'])
df_new.loc['worst_Xi']=(df_new.loc['Beneficial']*df_new.loc['Min'])
                      +(df_new.loc['Non-Beneficial']*df_new.loc['Max'])
df_new.loc['wts']=[0.049752,0.099672,0.164711,0.082165,0.170663,0.301525,0.131512]
df_extended['Si'] = np.sum((((df_new.loc['best_Xi'] - df) /
                    (df_new.loc['best_Xi'] - df_new.loc['worst_Xi']))*df_new.loc['wts']),axis=1)
df_extended['Ri'] = np.max((((df_new.loc['best_Xi'] - df) /
                    (df_new.loc['best_Xi'] - df_new.loc['worst_Xi']))*df_new.loc['wts']),axis=1)
s_best = np.min(df_extended['Si'])
r_best = np.min(df_extended['Ri'])
s_worst = np.max(df_extended['Si'])
r_worst = np.max(df_extended['Ri'])
u = 0.5
df_extended['Qi'] = ((((df_extended['Si'] - s_best) /(s_worst - s_best)) * u)
                    + ((1-u)*((df_extended['Ri'] - r_best)/(r_worst-r_best))))
df_extended['rank_based_on_Si'] = df_extended['Si'].rank(ascending=True)
df_extended['rank_based_on_Ri'] = df_extended['Ri'].rank(ascending=True)
df_extended['rank_based_on_Qi'] = df_extended['Qi'].rank(ascending=True)
QA1 = df_extended[df_extended.rank_based_on_Qi==1].loc[:,'Qi'][0]
QA2 = df_extended[df_extended.rank_based_on_Qi==2].loc[:,'Qi'][0]
DQ = 1/(len(df)-1)
Condition1 = False
Condition2 = False
if (QA2 - QA1) > DQ:
    Condition1 = True
```

Figure 7.6   *Continued*

```
if (df_extended[df_extended.rank_based_on_Qi==1].loc[:,'Qi'].index[0]
        == df_extended[df_extended.rank_based_on_Ri==1].loc[:,'Ri'].index[0])
or (df_extended[df_extended.rank_based_on_Qi==1].loc[:,'Qi'].index[0]
        == df_extended[df_extended.rank_based_on_Si==1].loc[:,'Si'].index[0]):
    Condition2 = True
if Condition1 and Condition2:
    print('both condition satisfied')
    print('best alternate solution is - '
        +df_extended[df_extended.rank_based_on_Qi==1].loc[:,'Qi'].index[0] )
else:
    if Condition1:
        print('only condition 2 is not satisfied. Best alternative solutions are:')
        print(df_extended[df_extended.rank_based_on_Qi==1].loc[:,'Qi'].index[0])
        print(df_extended[df_extended.rank_based_on_Qi==2].loc[:,'Qi'].index[0])
    if Condition2:
        print('only condition 1 is not satisfied. Best alternative solutions are:')
        soln_array = []
        # rank1 is always included in answer
        soln_array.append(df_extended[df_extended.rank_based_on_Qi==1].loc[:,'Qi'].index[0])
        for a in range(2,len(df)+1):
            if (df_extended[df_extended.rank_based_on_Qi==a].loc[:,'Qi'][0] - QA1) < DQ:
                soln_array.append(df_extended[df_extended.rank_based_on_Qi==a].loc[:,'Qi'].index[0])
            else:
                break
        for a in soln_array:
            print(a)

both condition satisfied
best alternate solution is - Collab Adjustment+ARIMA
```

Figure 7.6 Python code for VIKOR.

References

Mehdiyev, N., Enke, D., Fettke, P., and Loos, P. (2016). Evaluating forecasting methods by considering different accuracy measures. *Procedia Computer Science, 95*, 264–271.

Cryer, J. D., and Chan, K. S. (2008). *Time Series Analysis: with Applications in R*. Springer Science & Business Media.

Badulescu, Y., and Cheikhrouhou, N. (2018). Evaluation of forecasting approaches using hybrid multi-criteria decision-making models. In *Proceedings of International Conference on Time Series and Forecasting (ITISE 2018)* (No. CONFERENCE). 19–21 September 2018.

Zhu, X., Zhang, G., and Sun, B. (2019). A comprehensive literature review of the demand forecasting methods of emergency resources from the perspective of artificial intelligence. *Natural Hazards, 97*(1), 65–82.

Nerlove, M., and Diebold, F. X. (1990). Autoregressive and moving-average time-series processes. In *Time Series and Statistics,* 25–35. Palgrave Macmillan, London.

Nielsen, A. (2019). *Practical Time Series Analysis: Prediction with Statistics and Machine Learning.* " O'Reilly Media, Inc.".

Elmunim, N. A., Abdullah, M., Hasbi, A. M., and Bahari, S. A. (2015). Comparison of statistical Holt-Winter models for forecasting the ionospheric delay using GPS observations. *94.20. Cf; 91.10. Fc; 95.75. Wx.*

Razali, S. N. A. M., Rusiman, M. S., Zawawi, N. I., and Arbin, N. (2018, April). Forecasting of water consumptions expenditure using Holt-Winter's and ARIMA. In *Journal of Physics: Conference Series* (Vol. 995, No. 1, p. 012041). IOP Publishing.

Xu, X., Qi, Y., and Hua, Z. (2010). Forecasting demand of commodities after natural disasters. *Expert Systems with Applications, 37*(6), 4313–4317. htt ps://people.duke.edu/$\sim$rnau/411arim.htm

Zhang, G. P. (2003). Time series forecasting using a hybrid ARIMA and neural network model. *Neurocomputing, 50,* 159–175.

Alencar, D. B., Affonso, C. M., Oliveira, R. C., and Jose Filho, C. R. (2018). Hybrid approach combining SARIMA and neural networks for multi-step ahead wind speed forecasting in Brazil. *IEEE Access, 6,* pp. 55986–55994.

Saaty, R. W. (1987). The analytic hierarchy process—what it is and how it is used. *Mathematical Modelling, 9*(3–5), 161–176

Macharis, C., Springael, J., De Brucker, K., and Verbeke, A. (2004). PROMETHEE and AHP: The design of operational synergies in multicriteria analysis.: Strengthening PROMETHEE with ideas of AHP. *European Journal of Operational Research, 153*(2), 307–317.

Duran, O., and Aguilo, J. (2008). Computer-aided machine-tool selection based on a Fuzzy-AHP approach. *Expert Systems with Applications, 34*(3), 1787–1794.

Forman, E. H., and Gass, S. I. (2001). The analytic hierarchy process—an exposition. *Operations Research, 49*(4), 469–486.

de FSM Russo, R., and Camanho, R. (2015). Criteria in AHP: a systematic review of literature. *Procedia Computer Science, 55,* 1123–1132.

Anandan, V., and Uthra, G. (2017). Extension of TOPSIS using L1 Family of Distance Measures. *Advances in Fuzzy Mathematics, 12*(4), 897–908.

Chen, C. T. (2000). Extensions of the TOPSIS for group decision-making under fuzzy environment. *Fuzzy Sets and Systems, 114*(1), 1–9.

Opricovic, S. (1990, October). Programski paket VIKOR za visekriterijumsko kompromisno rangiranje. In *17th International Symposium on Operational Research SYM-OP-IS.*

Opricovic, S., and Tzeng, G. H. (2003). Defuzzification within a multicriteria decision model. *International Journal of Uncertainty, Fuzziness and Knowledge-Based Systems, 11*(05), 635–652.

Opricovic, S., and Tzeng, G. H. (2004). Compromise solution by MCDM methods: A comparative analysis of VIKOR and TOPSIS. *European Journal of Operational Research, 156*(2), 445–455.

Opricovic, S., and Tzeng, G. H. (2007). Extended VIKOR method in comparison with outranking methods. *European Journal of Operational Research, 178*(2), 514–529.

Sayadi, M. K., Heydari, M., and Shahanaghi, K. (2009). Extension of VIKOR method for decision making problem with interval numbers. *Applied Mathematical Modelling, 33*(5), 2257–2262.

8

Rainfall Prediction Using Artificial Neural Network

Sunil K. Sahu[1], N. Kumar Swamy[1], and Dinesh Bisht[2]

[1]School of Sciences, ISBM University Nawapara (Kosmi),
Block & Tehsil-Chhura, Gariyaband, Chhattisgarh-493996, India
[2]Jaypee Institute of Information Technology, Noida, India
E-mail: nkumarswamy15@gmail.com

Abstract

The multilayer artificial neural network in the back propagation algorithm is commonly used as a tool for rainfall prediction. The result of its prediction is reliable and precise. In this study, rainfall prediction in the Gariyaband region in the state of Chhattisgarh, India has been predicted by using an artificial neural network Backward Propagation Algorithm tool. Three-layer models are used to study various characteristics of a hidden neuron in a network. It is observed that in the present value as the number of neurons increases in an artificial neural network the mean square error decrease. The Back-Propagation algorithm is the leading algorithm for monsoon prediction.

Keywords: Back Propagation Algorithm, Artificial Neural Network, Prediction, Rainfall and Multilayer Artificial Neural Network.

8.1 Introduction

Predicting rainfall is a complex process in terms of reliability and precision is concerned. Predicting requires a large amount of storage and past real data in agriculture. Rainfall prediction plays a very important role in irrigation, increase production (Purnomo et al., 2017). Rain forecast helps water management and flood forecasting. Factors that influence rain forecast such as temperature, humidity, wind speed, pressure, dew point, etc (Enireddy et al., 2010) influence rain forecast.

To predict the desired rainfall from the acquired data, a variety of forecasting methods are available. Back propagation algorithm, layer recurrent network, and cascaded back propagation are the three most often used approaches used for the purpose. K.W. Wong et al. employed ANN to learn the relationship between the observed data in compassionate circumstances where human comprehension of physical processes is not clear (Fung et al., 1997).

The learning cycle of a back propagation algorithm comprises two phases: one for propagating the input pattern across the network and the other for adapting the output by modifying the weights in the network. Through a mechanism known as chain rule, the technique is utilized to successfully train a neural network. Back propagation, in simple terms, performs a backward pass over a network. After each forward pass, the cycle continues with model parameters (weights and biases).

Sharma et al. (Ankita Sharma and Geeta, 2015) implemented rainfall prediction by using an ANN propagation algorithm. It is suggested to use such a tool to have more precision in the predicted results. The purpose of the Back Propagation Algorithm is to overcome errors weights. ANN learns the training data for the purpose. Rainfall can be predicted by using the back propagation technique. The training, testing and detection of the hidden neuron in the network are covered in this study. Because of their ability to examine and determine the historical data needed for prediction. Hu et al. execution of Artificial Neural Network, considered as a significant delicate processing technique in climate anticipating (C and Hu, 1964). ANN has preferable precision over measurable and numerical models. Artificial neural network outperforms Statistical and mathematical model in terms of accuracy. ANN is based on the biological neuron principle.

Geeta et al. (Geeta and Samuel Selvary, 2011) developed rainfall prediction in Chennai city in the state of Tamil Nadu, India by using a back propagation neural network. An artificial neural network can be used to investigate objectives the relationship between metrological parameters and rainfall. The goal of the study is to create an ANN that uses back propagation to predict rainfall in the Gariyaband district of Chhattisgarh. Indian economy is a gamble of monsoon. Weather prediction is the tool to guide farmers for better production. The application of ANN can serve the agricultural community and water-deficient farmer community of Gariyaband district of Chhattisgarh, India.

8.2 Materials and Method

The process involved three sequential phases namely, input, output data selection, input data training, testing and validation.

8.2.1 Input and Output Data Selection

Monsoon rainfall data of Gariyaband District has been considered as input. From mid-June to the end of September the months of July and August are the most wetted. As a result, the current analysis looks at the data from 2009 through 2020 month by month. The output and input files have 144 entries. The average humidity and average wind speed in the last 12 years from 2009–2020 are the input parameters. It's now a 2×144 matrix. The input files have two rows and 144 columns, while the output files have one row and 144 columns are ascertained from the information www.worldweatheronline.com.

8.2.2 Input Data Training

The next step is to train the input data using the MATLAB back propagation algorithm after getting it (BPA). A three-layered ANN back propagation learning model is used in the study. Only 70% of the input data is used for training, therefore out of 144 samples, only 100 are used for training and these are chosen at random. For each attempt at training the data, the algorithm selects a training sample at random from the entire collection and a fixed set of data, resulting in a different mean square error value each time the data is trained (MSE). The remaining 22 samples are kept for testing, while the rest 22 samples are preserved for validation.

8.2.3 Validation and Testing

After the data has been trained and the error has been reduced to below the tolerance thresholds, testing is carried out. The BPA keeps 30% of the input data for testing and validation, which means that out of 144 samples, 22 are used for testing and another 22 for validation.

8.2.4 Artificial Neural Network Architecture

The structure and function of a biological neural network are used to design artificial neural network architecture. ANN is made up of neurons that are arranged in layers similar to neurons in the brain. The feed-forward is a popular neural network that has also three layers: An input layer receives the external date of pattern recognition whereas the output layer solves the problem. A hidden layer connects other levels. The artificial neural network learns the database by using a training algorithm that modifies the neuron's weights basing on the error rate of tested and actual output. The general structure of ANN is shown in Figure 8.1

Abishek et al. developed an artificial neural network model to predict the average rainfall in the Udupi district of state Karnataka, India. Average wind speed and humidity were used as input with average rainfall from 1960–2010 being the output. In the Matlab tools, three ANN models were used: feed-forward with back propagation, layer recurrent network and cascaded feed-forward back propagation. The anticipated values in the back propagation method have the same trend as the target value. When compared to the BPA the mean square error in the layer recurrent network was rather significant in several circumstances. When compared to the back propagation algorithm and layer recurrent networks the cascaded approach has a high mean square error in practically every instance. As a result among th three algorithms examined the back propagation algorithm is best.

(Abhishek et al., 2012). Debnath et al. developed an ANN model of predicting average summer-monsoon rainfall over north and south Assam in India has been analysed. The method used in three ANN models with sigmoid non-linearity. The ANN model is step by step to predict the average rainfall

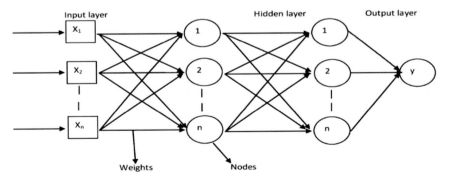

Figure 8.1 Architecture of artificial neural network.

over north and south Assam during the summer monsoon between the period 1871–2000. After analysed the data the result compares the output of the regression approach found that for south Assam, the neural network completely outperforms the regression. The ANN model with two inputs is found the best result. After observing two inputs ANN with multiple regressions in the case of north Assam produced a better forecast, but for south Assam the two inputs ANN is found to be the best predictive model for monsoon rainfall (De et al., 2009). Saranjeet et al. developed an ANN model for weather prediction of state Haryana, India. The maximum, minimum, relative humidity and average wind speed are the meteorological parameters. The most significant algorithm for weather forecasting is the back propagation method. The LM algorithm is used to train the data and it is the quickest technique of weather forecasting, although there are many different BPA has a higher learning rate (Pooja Malik et al., 2014).

TOOL in MATLAB was used to create a multilayer architecture. BPA, cascaded back propagation algorithm and layer recurrent network is all put to the test in multilayer architecture. Nirvesh et al. developed a model on rainfall prediction for the Chao Phraya River, Thailand using Neural Network with online data collection during the period from 2002–2005 record online 15 minutes all year round. They have used the back propagation neural network provided by Stuttgart neural network prediction of the rainfall data using back propagation technique acceptable with an accuracy of 97.42 and 95.44 for training and testing set respectively (Niravesh Srikalra and Chularat Tanprasert, 2006).

8.3 Result

In the BPA for multilayer architecture, three hidden layers used one output, one output layer and 10–20 neurons per layer were developed to be used in the experiment. The result of different cases of BPA is shown in Table 8.1 case-01 has the minimum MSE was 2.21 and the maximum MSE was 5.45. The case-01 MSE is the best case for the BPA it performances are plotted in Figure 8.2. The graph was built between the mean square error, the train, the verification and the test parameter design for the best-case scenario. The best validation check occurred at epoch 0. Figure 8.3 shown the comparison between actual and predicted values for the BPA and the regression plot was plotted in Figure 8.4. The outcome of the various LRN cases shown in Table 8.2 is all cases such as the BPA except the mean square error to be significantly different from

Table 8.1 Comparison of MSE for different cases using back propagation algorithm

Sl. No.	Training Function	Adaptive Learning Function	No of Neuron	MSE
Case 01	TRAINLM	Learngdm		2.21
Case 02		Learngd	10	3.45
Case 03		Learngdm	10	3.01
Case 04		Learngd	20	3.99
Case 05	TRAINRP	Learngdm	20	4.45
Case 06		Learngd		5.45
Case 07		Learngdm	10	2.99
Case 08		Learngd	10	3.05

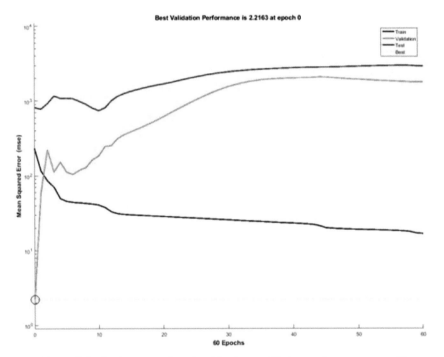

Figure 8.2 Back propagation algorithm best validation performances case-01.

that of the BPA. The best case in LRN was case-04 with the least MSE of 2.54. In LRN the performances of case-04 shown in Figure 8.5, the best validation performances are accrued at epoch 0. Figure 8.6 shows the comparison between Actual and predicted values for LRN. The result of various cases in Cascaded back propagation shown in Table 8.3 the best case of CBP algorithm is case-03 with the least MSE 5.35 the performances of

8.3 Result 133

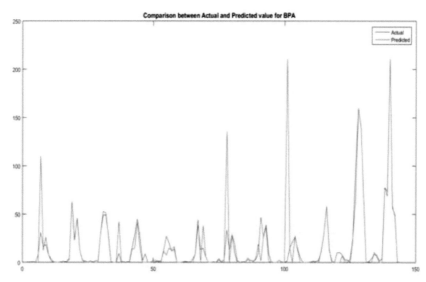

Figure 8.3 Comparison between actual and predicted value for back propagation algorithm.

Table 8.2 Comparison of MSE for different cases for layer recurrent network

Sl. No.	Training Function	Adaptive Learning Function	No of Neuron	MSE
Case 01	TRAINLM	Learngdm		8.99
Case 02		Learngd	10	7.35
Case 03		Learngdm	10	5.35
Case 04		Learngd	20	2.54
Case 05	TRAINRP	Learngdm	20	6.28
Case 06		Learngd		4.38
Case 07		Learngdm	10	3.97
Case 08		Learngd	10	3.00

CBP case-03 validation efficiency was obtained epoch 0 and the regression is shown in Figure 8.8. The graph is built between actual and predicted values for CBP. The CBP algorithm showed a good level of accuracy. In Figure 8.9 shows a comparison between actual and predicted values for CBP.

Figure 8.10 shows the average prediction error corresponding to the BPA, CBP, and Layer recurrent network based on neural network predictive model. The result shows that the BPA produced the lowest prediction error among the three models. In Figure 8.11 shows the prediction error from the BPA. It is found that in 75 out of 144 sample cases the prediction error lies below 0.5. it can, therefore, conclude that the BPA produced a significant forecast yield.

134 Rainfall Prediction Using Artificial Neural Network

Table 8.3 Comparison of MSE for different cascaded back propagation

Sl. No.	Training Function	Adaptive Learning Function	No of Neuron	MSE
Case 01	TRAINLM	Learngdm	10	6.25
Case 02		Learngd	10	6.90
Case 03		Learngdm	20	5.35
Case 04		Learngd	20	7.35
Case 05	TRAINRP	Learngdm	10	6.00
Case 06		Learngd	10	5.90
Case 07		Learngdm	20	6.17
Case 08		Learngd	20	7.35

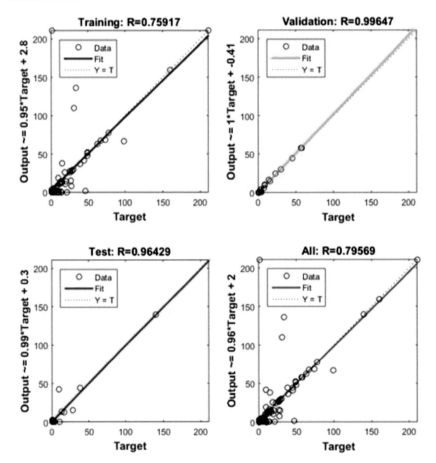

Figure 8.4 Regression plot in back propagation algorithm.

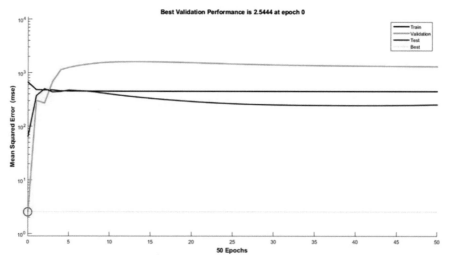

Figure 8.5 Layer recurrent network best validation performances case-04.

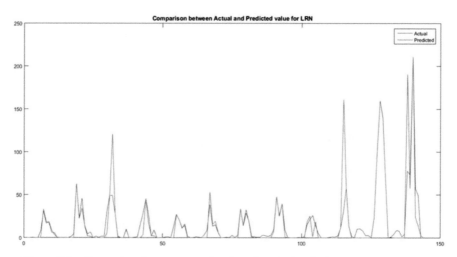

Figure 8.6 Comparison between actual and predicted value for layer recurrent network.

8.4 Discussion

The BPA has a high-level accuracy and the projected value follows the same trend as the target value, with the least regression plot divergence in the graph, but the predicted values in some cases. An increasing number of neurons in the MSE was decreased. While the layer recurrent network has a good level of

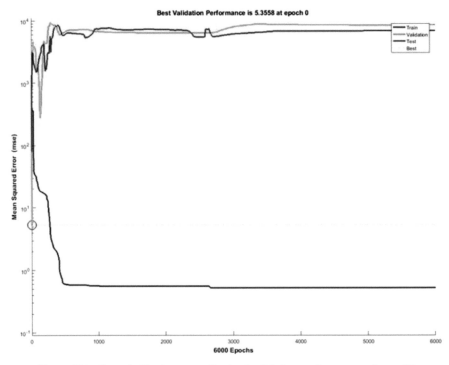

Figure 8.7 Cascaded back propagation best validation performances of case-03.

accuracy, the MSE was rather high in several circumstances when compared to BPA. In comparison to the BPA and layer recurrent network, the cascaded displayed a higher MSE. The CBP the predicted value goes same trend and better accuracy. The mean square error of artificial neural networks lowers as the number of increases, LEARNGD takes a little longer and the higher the amount of input data, the lower is in mean square error after training.

8.5 Comparision of ANN Model with Regresion

In the earlier section, it is established that the BPA can be considered as a good predictive methodology. In this section, the prediction from the BPA model would be placed against the multiple linear regressions to have a comparative study.

A multiple linear regression is of the form

$$y = b_0 + b_1 x_1 + b_2 x_2 \tag{8.1}$$

8.5 Comparision of ANN Model with Regresion 137

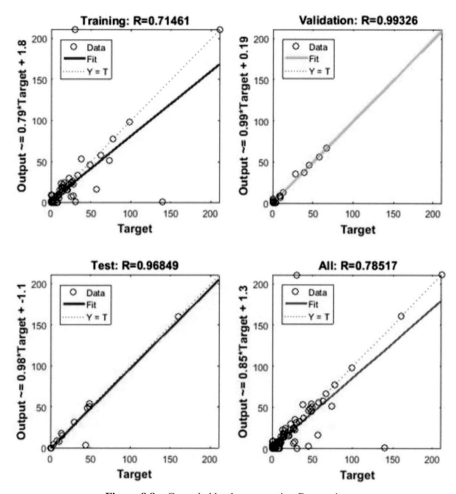

Figure 8.8 Cascaded back propagation Regression.

Where bi represents the regression coefficient, x1 represents the wind speed and x2 represents humidity. In multiple linear regression, the average wind speed and average humidity are taken as a predictor and the average rainfall values are the predictand. The prediction error from Multilayer linear regression is plotted against the BPA. The Figure 8.12 shows the multiple linear regression out performances by the Back propagation algorithm. Thus, it was found that for BPA also produced better prediction than multiple linear regressions.

138 Rainfall Prediction Using Artificial Neural Network

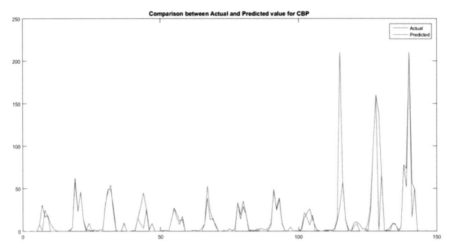

Figure 8.9 Comparison between actual and predicted value for cascaded back propagation

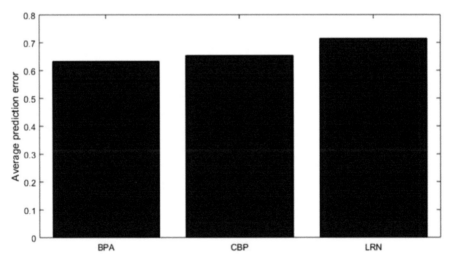

Figure 8.10 Average prediction error corresponding to back propagation algorithm, cascaded propagation algorithm and layer recurrent network.

8.5 Comparision of ANN Model with Regresion 139

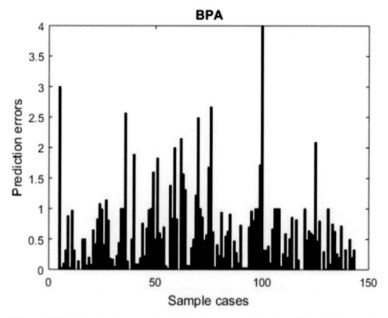

Figure 8.11 Prediction error for the back propagation algorithm ANN model.

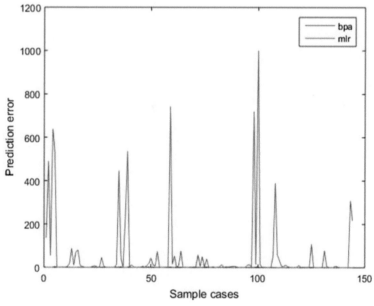

Figure 8.12 Prediction errors from back propagation algorithm and multiple linear regression.

8.6 Conclusion

For rainfall prediction, an artificial neural network was applied to predict rainfall data of Gariyaband District of state Chhattisgarh, India. According to the experiment, the prediction of rainfall data using the BPA was acceptable with the least MSE of 2.21. The BPA is best to compare to all the three tested. After comparing the BPA with multiple linear regression, it was observed that the ANN model BPA produced a better forecast.

In future work, additional input can be used for rainfall prediction inputs such as temperature, pressure, dew point, etc.

References

Purnomo, H. D., Hartomo, K. D., Prasetyo, S. Y. J. (2017). Artificial neural network for monthly rainfall rate prediction. *IOP Conf. Series: Material Science and Engineering.*

Enireddy, V., Varma, K. V. S. R., Sankara Rao, P., Ravikanth, S. (2010). Prediction of rainfall using back propagation neural network model. *International Journal of Computer Science and Engineering,* 02(04), 1119–1121.

Fung, C. C., Wong, K. W., Eren, H., Charlebois, R., Crocker, H. (1997). Modular artificial neural network for prediction of petrophysical properties from well long data. *IEEE Transaction on Instrumentation and Measurement,* 46(6), 1259–1263.

Sharma, A., Geeta, N. (2015). Rainfall prediction using neural network, *International Journal of Computer Science Trends and Technology*, 3(3), 65–69.

C, M. J., and Hu. (1964). Application of ADALINE system to weather forecasting technical report, *Stanford Electron.*

Geeta, G., Samuel Selvary, R. (2011). Prediction of monthly rainfall in Chennai using back propagation neural network model, *International Journal of Engineering Sciences and Technology,* 3(1), 0975–5462.

Abhishek, K., Abhay Kumar, R. Ranjan, and Kumar, S. (2012). A rainfall prediction model using artificial neural network, *IEE Control and System Graduate Research Colloquim*, 82–87.

De, S. S., G. C., and A. Debnath. (2009). ANN- a tool for prediction monsoon rainfall over north and south Assam in India, *Bulg. J. Phys.* 36, 55–67.

Pooja Malik, Saranjeet Singh, and Binni Arroa. (2014). An effective weather forecasting using neural network, *IJEE, Research and Technology,* 02(02), 209–212.

Niravesh Srikalra and Chularat Tanprasert. (2006). Rainfall prediction for Chao Phraya River using neural network with online data collection. Proceedings of the 2nd IMT-GT Regional conference on Mathematics, Statistics and Applications University Saints Malaysia, Penang.

www.worldweatheronline.com/gariaband-weather-averages/chhattisgarh/in.aspx

9

Statistical Downscaling and Time Series Analysis for Future Scenarios of Temperature in Haridwar District, Uttarakhand

Gaurav Singh[1], Nitin Mishra[2], and L. N. Thakural[3]

[1]Punjab Remote Sensing Centre, Ludhiana, India
[2]Assistant Professor, Department of Civil Engineering, Graphic Era Deemed to be University, Dehradun, India
[3]Water Resources Division, National Institute of Hydrology, Roorkee, India
E-mail: gaurav.panwar.gs@gmail.com; nitinuag@gmail.com; thakuralln@gmail.com

Abstract

Environmental variation can cause large effects on water assets by resulting in variations in the hydrological cycle. Temperature is one of the most significant meteorological variables since it tends to be identified with sunlight-based radiation and accordingly with both evaporation and transpiration forms which establish a significant period of the hydrologic cycle. Time series prediction is the cycle of cautious assortment and thorough investigation of information that has been gathered over a nonstop period and improvement of a legitimate model that portrays the innate pattern of the arrangement. The examination considers completed by different scientists found that the expanding patterns in air temperature have been identified with a few factors, for example, expanded concentrations of anthropogenic greenhouse gases, expanded emanations of anthropogenic aerosols, expanded cloud cover and urbanization. GCMs (General circulation models) in addition to large-scale circulation predictors are considered as the two most critical and significant means to study the environmental effects. Statistical Downscaling technique utilized for downscaling daily temperature data to obtain future climate data for the Haridwar district in Uttarakhand. The large-scale NCEP (National Centers for Environmental Prediction) 'reanalysis' data of the interval 1961–1995, utilized for calibration and the data of the interval 1996–2005, utilized for the model's validation. The estimation of future monthly based temperature for the period 2020s, 2050s and 2080s for Haridwar district is carried out for different RCPs (2.6, 4.5 and 8.5). Both Average Annual maximum temperature and Average Annual minimum temperature shows an increasing trend for 2020s, 2050s and 2080s for all three scenarios.

Keywords: Climate change, NCEP Predictors, Time Series Analysis, Future Scenarios, SDSM, Rainfall, RCP, Haridwar district.

9.1 Introduction

Time series analysis has made some amazing progress directly since its commencement. A lot of examination has been done in time arrangement investigation to achieve different targets. Time series forecast is basically a segment of temporal data mining and insights. This can be seen as the interaction of cautious assortment and thorough investigation of information that has been gathered over a persistent interval and advancement of an appropriate model which portrays the inborn pattern of the arrangement (Athiyarath et al., 2020). With respect to hydrology, variation in the environment can have significant impacts on water resources by causing variations in the hydrologic cycle. Temperature and precipitation are the fundamental factors, which impacts legitimately cause environmental change (Hassan and Harun, 2011). Temperature is one of the most significant meteorological variables since it tends to be identified with sunlight-based radiation and accordingly with both transpiration and evaporation forms which establish a significant period of the hydrologic cycle. The IPCC (2007) has reported on increasing temperature trends in many areas of the world in time and space. The examination considers completed by different scientists found that the expanding patterns in air temperature have been identified with a few factors, for example, expanded concentrations of anthropogenic greenhouse gases, expanded emanations of anthropogenic aerosols, expanded cloud cover and urbanization (Duhan et al., 2011). Outrageous climate occasions are probably going to cause harm to the environment and society. The strength of expansion as well as the recurrence of such opportunities is of extraordinary concern and by various reviews on the future variation of the environment, the validation is carried out throughout the world. An increase in temperature and precipitation limits will result in a lengthening of dry seasons and may result in an extension of areas influenced by the dry period (Fischer et al., 2011). Different RCPs (representative concentration pathways) length the scope of conceivable radiative forcing situations and are used by the Climate modeling community for concentrations of aerosol and greenhouse gas together with changes in land use, that are steady with a lot of wide atmosphere results. RCP 2.6 represents a very less forcing level while RCP4.5 and RCP6 represent moderate stabilization situations and RCP 8.5 depicts a situation of very high baseline radiation (Jubb et al., 2013). The future environmental variation due to consistent increment in the amount of greenhouse gases in the air can be estimated with the help of GCMs (General circulation models) (Shukla et al., 2015). GCMs in addition to large-scale circulation predictors consider as two most critical and important means of addressing environmental effects (Yang et al., 2011). GCM helps predict future meteorological parameters under various conditions for future climate scenarios. Because GCMs have a coarser spatial resolution, they cannot use for local impact assessments. As a result, to obtain local-scale surface weather from regional-scale atmospheric predictor variables downscaling is used. Statistical downscaling derives statistical relationships between local-scale predictands and regional scale predictors. (Goyal and Ojha, 2010). The selection of predictors shows its nature, correlation, dependency and principal components (Wilby and Wigley, 2000).

Downscaling strategies are usually categorized as statistical and dynamic downscaling. The technique of statistical downscaling builds a statistical connection between nearby climate factors and large-scale GCM yields and the method of dynamic downscaling utilizes high-resolution local atmosphere models settled in a GCM for acquiring neighborhood climate factors. In the hydrology study, it has been reported that statistical downscaling, generally considered less computational (Chen et al., 2010).

9.2 Study Area

Haridwar district, situated in the state Uttarakhand, India. This place is for nature lovers is considered a holy place of Hinduism. As per the 2011 census, Haridwar district is the most populous district of Uttarakhand. The latitudes

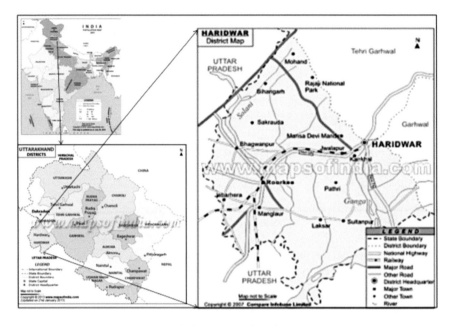

Fig. 1. Location map of study area

Figure 9.1 Map of study area (source: www.mapsofindia.com).

and longitudes of Haridwar district are 29°33' 30°14' N and 77°57'–78°1' E respectively. The geographical area of Haridwar district is approximate 2360 km^2, which comprises of six legislative blocks viz. Roorkee, Bhagwanpur, Narsan, Laksar, Khanpur and Bahadrabad (Figure 9.1). The mean temperature of Haridwar district is 22°C and normal rainfall in the haridwar district is about 1134 mm in which most of the rainfall occurs in the monsoon season (June–September).

9.3 Data Used and Methodology

9.3.1 Data Used

'Reanalysis' is a project contributed by the National Center for Atmospheric Research (NCAR) and NCEP together to deliver documentation of 50 years of worldwide investigation of climatic information as air. It is then prepared for standard check and incorporated with a complicated & superior information integration framework that is kept unaltered over the whole 'reanalysis' period. The 'Reanalysis' data, which, considered as an intermediary to observed information, has $2.5° \times 2.5°$ of resolution (Kistler et al., 2000). In this study, 45 years of the predictor data (reanalysis data) were observed on a daily basis of atmospheric variables used for the period of 1961–2005 from the Canadian Climate Impacts Scenarios (CCIS) website (www.cics.uvic.ca/scenarios/sdsm/select.cgi).

The GCM daily temperature simulations acquired from Coupled Model Inter-comparison Project 5 (CMIP5) climate change investigations for the period 2006–2100. CanESM2 model output data used in this study and affiliated to the Canadian Centre for Climate Modelling and Analysis (CCCMA). The information is available online.

The statistics of temperature ($1° \times 1°$) on a daily basis for the period of 1961–2005 were obtained from the Indian Meteorological Department (IMD). Statistical downscaling has been performed utilizing the day-to-day precipitation time arrangement as input predictand in SDSM software.

9.3.2 Methodology

SDSM (Statistical Downscaling Model) works on multiple regression techniques to obtain future situations to evaluate the effects of environmental change. Two types of data are required in this model, i.e. 1. the local data as 'Predictand' (temperature) and 2. various atmospheric variables known as 'Predictors'. In this study, Multiple Linear Regression strategy (parametric) has been utilized. SDSM tool version 4.2 is used for downscaling (Chandniha and Kansal, 2011).

Appropriate predictors were chosen using both positive and negative correlation and partial correlation analysis between the predictors and the temperature (predictand). Calibration of the model has been performed to establish an empirical relation between the predictors and the predictand using the multiple linear regression (parametric) techniques. Future scenarios for different RCPs emissions for the selected study area utilizing the simulated CanESM2 GCMs data generated based on the validated regression model. The study assumed that for the future climate conditions, this empirical relation between predictor variables and predictand (temperature) remains valid.

9.4 Results and Discussion

A time series is non-deterministic in nature, for example, we can't anticipate with certainty what will happen in the future. Generally, a time series $\{x(t)t = 0, 1, 2, \ldots\}$ is accepted to follow a certain probability model which depicts the joint distribution of the irregular variable X_t (Adhikari and Agrawal, 2013).

9.4.1 Regression Method

Regression strategies are the most broadly utilized and simple to execute techniques when a relationship has to be drawn between the anticipated value and different components. An assumption is that, the information to be anticipated is separated into linear data and standard base information is reliant upon the components that are influencing the information. For instance, in the case of forecasting electrical load, the load relies upon the accompanying variables like the customer qualities, months, weekdays and conditions of weather. A matrix (Xt) is an inherent direct model that incorporates past and present observations of indicators in order by time (t). This technique multiple linear regression model (MLR) has been utilized $y(t) = x(t)b + u(t)$ to obtain the assessment of a linear connection of reaction $y(t)$. In this β goes about as a linear parameter and $u(t)$ addresses innovation term. The matrix $x(t)$ is characterized dependent on the variables, based on which time series data vary, for instance, local time, months, climate and so forth (Athiyarath et al., 2020).

9.4.2 Predictor Variables Selection

A number of 26 predictor variables of a large-scale are considered during initial screening. The predictors utilized are dependent on the relationship and the partial connection of NCEP's predictors and observed environment factors in a downscaling model for the interval 1961–2005. Tables 9.1 and 9.2 show the selected predictors using the correlation coefficients values, partial correlation values and p values between predictand (Max. temperature and Min. temperature) and NCEP predictors.

9.4.3 Calibration and Validation Results

NCEP 'reanalysis' data of the interval 1961–1995 was utilized for calibrating the model and the data of the interval 1996–2005 was used for model validation. The SDSM model represents a good relationship between the observed and estimated monthly average statistics of temperature. The Value of Coefficient of determination (R^2) used during calibration and validation. Observed and estimated data during calibration (1961–1995) and

9.4 Results and Discussion

Table 9.1 Selected NCEP predictors and their relationship with maximum temperature

S.No	Selected Predictors	Description	Correlation Coefficients	Partial r	P value
1	ncepp500gl.dat	500 hpa geopotential height	0.730	0.013	0.1471
2	ncepmslpgl.dat	Mean sea level pressure	−0.724	−0.314	0.0000
3	ncepp5_fgl.dat	500 hpa airflow strength	−0.553	−0.096	0.0000
4	ncepp5_ugl.dat	500 hpa zonal velocity	−0.508	0.130	0.0000
5	ncepp850gl.dat	850 hpa geopotential height	−0.522	0.335	0.0000
6	nceptempgl.dat	Mean temperature at 2 m	0.891	0.209	0.0000

Table 9.2 Selected NCEP predictors and their relationship with minimum temperature

S.No	Selected Predictors	Description	Correlation Coefficients	Partial r	P value
1	ncepmslpgl.dat	Mean sea level pressure	−0.864	−0.192	0.0000
2	ncepp500gl.dat	500 hpa geopotential height	0.807	0.242	0.0000
3	ncepp5_ugl.dat	500 hpa zonal velocity	−0.714	−0.067	0.0000
4	nceps850gl.dat	Specific humidity at 850 hpa	0.782	−0.235	0.0000
5	ncepshumgl.dat	Surface specific humidity	0.746	0.314	0.0000
6	nceptempgl.dat	Mean temperature at 2 m	0.935	0.577	0.0000

validation (1996–2005) for max. temperature and min. temperature showed in Figure 9.2(a, b), and Figure 9.3(a, b) respectively.

Results of the observed and modeled monthly avg. data for both calibration and validation presented in Tables 9.3 and 9.4.

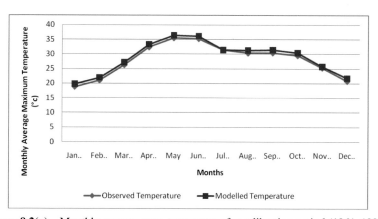

Figure 9.2(a) Monthly average max. temperature for calibration period (1961–1995).

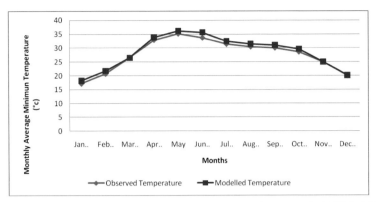

Figure 9.2(b)   Monthly average max. temperature for validation period (1996–2005).

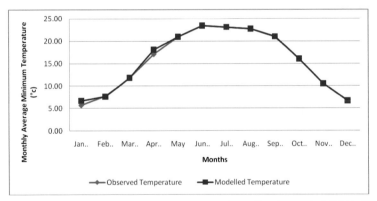

Figure 9.3(a)   Monthly average min. temperature for calibration period (1961–1995).

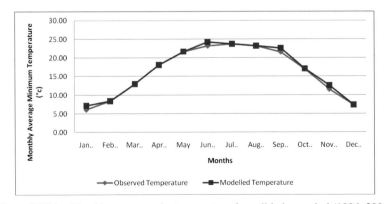

Figure 9.3(b)   Monthly average min. temperature for validation period (1996–2005).

9.4 Results and Discussion

Table 9.3 Calibration and validation of monthly average maximum temperature (°C) with NCEP reanalysis data months calibration

Months	Calibration (1961–1995)		Validation (1996–2005)	
	Observed	Modelled	Observed	Modelled
January	18.75	19.76	17.19	18.16
February	20.95	21.96	20.70	21.67
March	26.18	27.19	26.53	26.44
April	32.28	33.29	32.86	33.83
May	35.42	36.39	35.11	36.12
June	35.18	36.17	33.65	35.60
July	31.39	31.41	31.33	32.34
August	30.28	31.27	30.37	31.38
September	30.39	31.38	29.97	30.97
October	29.44	30.44	28.53	29.55
November	25.26	25.70	24.91	24.92
December	20.73	21.73	20.00	20.02
Minimum	18.75	19.76	17.19	18.16
Maximum	35.42	36.39	35.11	36.12
R^2	0.996		0.993	

Table 9.4 Calibration and validation of monthly average minimum temperature (°C) with NCEP reanalysis data months calibration

Months	Calibration (1961–1995)		Validation (1996–2006)	
	Observed	Modelled	Observed	Modelled
January	5.71	6.73	6.09	7.13
February	7.69	7.69	8.40	8.37
March	11.88	11.88	12.98	12.96
April	17.12	18.11	18.11	18.10
May	20.98	20.98	21.68	21.67
June	23.45	23.44	23.19	24.20
July	23.10	23.10	23.72	23.70
August	22.70	22.70	23.15	23.16
September	21.02	21.03	21.57	22.58
October	16.07	16.06	16.99	17.05
November	10.57	10.55	11.49	12.50
December	6.70	6.69	7.34	7.36
Minimum	5.71	6.69	6.09	7.13
Maximum	23.45	23.44	23.72	24.20
R^2	0.996		0.994	

9.4.4 Future Emission Scenarios

The future scenario generation of temperature under different Can ESM2 RCPs (2.6, 4.5 and 8.5) emission situations approximated with the identified predictors by MLR based SDSM model. The estimated temperature is represented into three different time steps, i.e. 2006–2040 (2020s) 2041–2070 (2050s) and 2071–2099 (2080s) for both Average Max. Temperature and Average Min. Temperature showed in Figure 9.4 (a, b, c) and 9.5 (a, b, c) respectively.

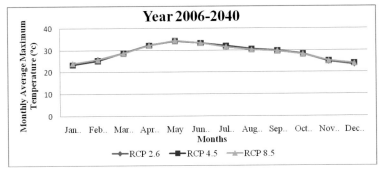

Figure 9.4(a)   Monthly average maximum temperature for climate scenarios 2006–2040.

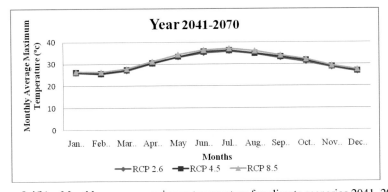

Figure 9.4(b)   Monthly average maximum temperature for climate scenarios 2041–2070.

9.4 Results and Discussion 153

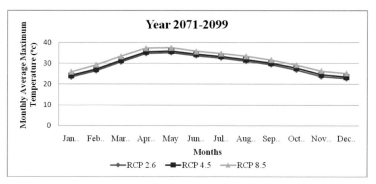

Figure 9.4(c) Monthly average maximum temperature for climate scenarios 2071–2099.

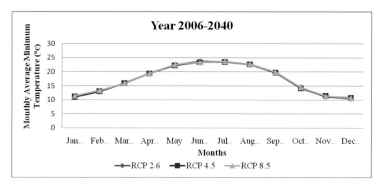

Figure 9.5(a) Monthly average minimum temperature for climate scenarios 2006–2040.

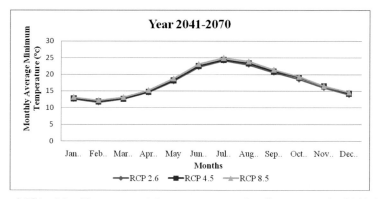

Figure 9.5(b) Monthly average minimum temperature for climate scenarios 2041–2070.

154 *Statistical Downscaling and Time Series Analysis for Future Scenarios*

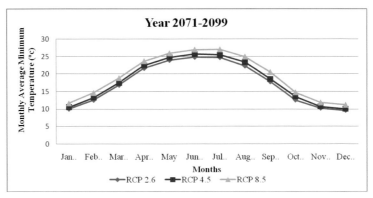

Figure 9.5(c) Monthly average minimum temperature for climate scenarios 2071–2099.

In Table 9.5, Under RCP 2.6 yearly maximum temperature ranges from 27.58°–30.48°C in 2020s, 29.75°–31.51°C in 2050s and 28.60°–30.46°C in 2080s. Similarly, under RCP 4.5, temperature varies from 27.84°–30.21°C in 2020s, 29.77°–31.64°C in 2050s and 29.39°–30.95°C in 2080s. Under RCP 8.5, temperature varies from 27.69°–30.49°C in 2020s, 30.21°–32.66°C in 2050s and 30.82°–32.70°C in 2080s.

Table 9.5 Detailed average maximum temperature (°C) statistics for different time steps (scenarios)

Month	RCP 2.6			RCP 4.5			RCP 8.5		
	2020's	2050's	2080's	2020's	2050's	2080's	2020's	2050's	2080's
January	23.65	26.02	23.39	23.42	26.14	24.18	24.05	26.52	25.90
February	25.87	25.46	26.37	25.28	25.60	27.27	25.78	26.36	29.30
March	28.76	26.91	30.62	28.67	27.27	31.44	28.67	27.83	33.32
April	32.40	30.20	34.71	32.21	30.38	35.54	32.30	31.18	37.25
May	34.24	33.13	35.16	34.23	33.33	35.89	34.50	34.39	37.41
June	33.27	35.09	33.71	33.30	35.63	34.39	33.31	36.40	35.70
July	31.49	35.88	32.63	32.04	36.17	33.36	31.60	37.13	34.70
August	30.27	34.86	31.01	30.56	34.92	31.78	30.23	36.02	33.43
September	29.70	32.82	29.40	29.74	33.26	30.10	29.76	33.99	31.67
October	28.15	31.00	26.82	28.28	31.56	27.78	28.32	32.12	29.23
November	24.87	28.51	23.59	25.16	28.81	24.61	25.22	29.30	26.38
December	23.47	26.56	22.68	24.02	26.81	23.52	24.00	27.52	25.21
Minimum	27.58	29.75	28.60	27.84	29.77	29.39	27.69	30.21	30.82
Maximum	30.48	31.51	30.46	30.21	31.64	30.95	30.49	32.66	32.70

Table 9.6 Detailed average minimum temperature (°C) statistics for different time steps (scenarios)

Month	RCP 2.6			RCP 4.5			RCP 8.5		
	2020s	2050s	2080s	2020s	2050s	2080s	2020s	2050s	2080s
January	11.04	12.59	10.05	10.98	12.71	10.54	11.41	13.13	11.67
February	13.22	14.58	12.59	12.92	11.75	13.22	13.32	12.20	14.63
March	15.90	12.63	16.83	16.00	12.57	17.45	15.95	13.09	18.85
April	19.54	14.56	21.64	19.40	14.69	22.42	19.46	15.24	23.71
May	22.49	17.89	23.98	22.25	18.11	24.74	22.57	18.75	25.97
June	23.76	22.19	24.84	23.38	22.53	25.67	23.36	23.15	26.99
July	23.51	24.17	24.74	23.49	24.40	25.43	23.72	25.00	27.06
August	22.58	22.95	22.32	22.69	23.34	23.41	22.78	23.95	25.00
September	19.75	20.55	17.81	19.84	20.93	18.55	19.95	21.48	20.54
October	14.57	18.46	12.63	14.19	19.02	13.61	14.53	19.31	14.79
November	11.16	15.99	10.30	11.24	16.41	10.71	11.37	16.50	11.90
December	10.41	13.89	9.64	10.79	14.19	10.07	10.71	14.59	11.22
Minimum	15.85	16.40	16.55	16.09	16.39	16.77	16.15	17.57	17.57
Maximum	18.46	18.45	18.00	18.68	18.32	19.59	18.86	18.54	21.39

In Table 9.6, Under RCP 2.6 yearly minimum temperature ranges from 15.85°C–18.46°C in 2020s, 16.40°C–18.45°C in 2050s and 16.55°C–18.00°C in 2080s. Similarly, under RCP4.5, temperature varies from 16.09°C–18.68°C in 2020s, 16.39°C–18.32°C in 2050s and 16.77°C–19.59°C in 2080s. Under RCP 8.5, 16.15°C–18.86°C in 2020s, 17.57°C–18.54°C in 2050s and 17.57°C–21.39°C in 2080s.

Tables 9.7 and 9.8 show an increasing trend for both Average yearly maximum temperature and average yearly minimum temperature compared with present scenario.

Table 9.7 Average yearly maximum temperature for present and downscaled maximum temperature corresponding to RCPs (2.6, 4.5 and 8.5) scenario

Interval	RCP 2.6	RCP 4.5	RCP 8.5
	Avg. Yearly Maximum Temperature (°C)		
Present (1961–2005)		27.95	
2020s (2006–2040)	28.85	28.92	28.99
2050s (2041–2070)	30.56	30.85	31.59
2080s (2071–2099)	29.18	29.99	31.63

Table 9.8 Average yearly minimum temperature for present and downscaled minimum temperature corresponding to RCPs (2.6, 4.5 and 8.5) scenario

Interval	RCP 2.6	RCP 4.5	RCP 8.5
	Avg. Yearly Minimum Temperature (°C)		
Present (1961–2005)		15.75	
2020s (2006–2040)	17.34	17.28	17.44
2050s (2041–2070)	17.32	17.58	18.06
2080s (2071–2099)	17.30	18.00	19.38

9.5 Conclusion

This study highlights the use of MLR based SDSM technique for assessing the likely future monthly Temperature variation in the Haridwar district. This study involves the generation of the daily temperature time series corresponding to a different RCPs (2.6, 4.5 and 8.5) scenarios and then utilized for estimating the monthly temperature for different future time. Average yearly maximum temperature exhibits an expanding pattern for 2020s, 2050s and 2080s for all the three scenarios. The average yearly minimum temperature also shows increasing trend for 2020s, 2050s and 2080s for all three scenarios. It is hoped that this the study would help in study the impact of climate variation on the temperature variation in this particular area.

References

Athiyarath S., Paul M. and Krishnaswamy S. (2020). A comparative study and analysis of time series forecasting techniques. *SN Computer Science*, 1:175.

Hassan Z.B., Harun S.B. (2011). Statistical downscaling for climate change scenarios of rainfall and temperature. *United Kingdom Malaysia Ireland Engineering Science Conference*, Kuala Lumpur 12-14 July 2011.

Duhan D., Pandey A., Gahalaut K.P.S. and Pandey R.P. (2013). Spatial and temporal variability in maximum, minimum and mean air temperatures at Madhya Pradesh in central India. C. R. *Geoscience* 345, 3-2.

Fischer T., Gemmer M., Luliu L. and Buda S. (2011). Temperature and precipitation trends and dryness/wetness pattern in the Zhujiang River Basin, South China, 1961-2007. *Quaternary International* 244, 138–148.

Jubb I., Canadell P. and Dix M. (2013). Representative Concentration Pathways (RCPs). Australian Climate Change Science Program.

Shukla R., Khare D. and Deo R. (2015). Statistical downscaling of climate change scenarios of rainfall and temperature over Indira Sagar Canal Command area in Madhya Pradesh, India. *IEEE 14th International Conference on Machine Learning and Applications.*

Yang T., Li H., Wang W., Xu C.Y. and Yu Z. (2011). Statistical downscaling of extreme daily precipitation, evaporation & temperature and construction of future scenarios. *Hydrological Processes.*

Goyal M.K. and Ojha C.S. (2010). Evaluation of various linear regression methods for downscaling of mean monthly precipitation in Arid Pichola Watershed. *Natural Resources.* 11–18.

Wilby R.L. and Wigley T.M.L. (2000). Precipitation predictors for downscaling: Observed and General Circulation Model Relationships. *International Journals.* 20, 641–661.

Chen S.T., Yu P.S. and Tang Y.H. (2010). Statistical downscaling of daily precipitation using support vector machines and multivariate analysis. *Journal of Hydrology* 385, 13–22.

Kistler R., Kalnay E., Collins W., Saha S., White G., Woollen J., Chelliah M., Ebisuzaki W., Kanamitsu M., Kousky V., Dool H.V., Jenne R. and Michael Fiorino (2000). *The NCEP–NCAR 50-Year Reanalysis: Monthly Means CD-ROM and Documentation.* American Meteorological Society.

Chandniha S.K. and Kansal M.L. (2016). Rainfall estimation using multiple linear regression based statistical downscaling for Piperiya watershed in Chhattisgarh. *Journal of Agro Meteorology* 18 (1), 106–112.

Adhikari R. and Agrawal R.K. (2013). An Introductory Study on Time Series Modeling and Forecasting.

Index

A
ACF 66, 71, 72, 73
AHP 100, 101, 108, 110, 115
Analysis 19, 21, 65, 143
Arima 65, 74, 105
Artificial neural network 81, 83, 127, 130

B
Back propagation algorithm 127, 128, 130, 132

C
Climate change 147
Coefficient of variation 1, 3, 6, 13
Commodity index 65, 69, 70, 75
Condition-based maintenance (CBM) 37, 38
Convolution neural network (CNN) 37, 38

D
Datasets 19, 21, 24, 120
Dimensionless indices 1
Divergence measure 2, 23
Dynamical changes 28, 34, 53, 56

E
Error measurement 99, 100, 103, 114, 120

F
Forecasting 68, 82, 99, 100, 102
Future scenarios 143, 147

H
Haridwar district 143, 145, 156
Hydrograph and hysteresis 82

I
Index 6, 20, 33, 65, 70

M
MCDM 99, 102, 114, 120
Modeling 42, 67, 82, 90

N
NCEP predictors 148, 149
Neural network 37, 68, 81, 83, 127
Nonlinear analyses 53

P
PACF 68, 73
Prediction 40, 65, 67, 82, 127, 139
Prognostic health management (PHM) 38

R
Rainfall and multilayer artificial neural network 127
Rating curve 82, 88, 89
RCP 144, 145, 154

Recurrence plots 37, 45
Runoff 82, 83, 88, 89

S

SDSM rainfall 144
Similarity measures 1, 4, 12, 21
Stationarity 70, 74
Suspended sediment 81, 82, 83, 86, 91

T

Time series 3, 5, 7, 19, 65, 93, 144
Time series analysis 1, 19, 102, 143
Topsis 101, 110, 116

V

Vikor 101, 102, 108, 112, 113, 119,

About the Editors

Dr. Dinesh C. S. Bisht received his Ph.D. with a major in Mathematics and a minor in Electronics and Communication Engineering from G. B. Pant University of Agriculture & Technology, Uttarakhand. Before joining the Jaypee Institute of Information Technology, he worked as an Assistant Professor at ITM University, Gurgaon, India. He has been a Faculty Member for around 11 years and has taught several core courses in Applied Mathematics and Soft Computing at undergraduate and master levels. The major research interests of him include Soft Computing and Nature Inspired Optimization. He has published more than 40 research papers in national and international journals of repute. He is the *Associate Editor* for *International Journal of Mathematical, Engineering and Management Sciences,* ESCI and SCOPUS indexed journal. He is the editor of the book "Computational Intelligence: Theoretical Advances and Advanced Applications" published by Walter de Gruyter GmbH & Co KG. He has also published seven book chapters in the reputed book series. Dr. Bisht is a member of International Association of Engineers in Hong Kong and Soft Computing Research Society, India. He has been awarded for outstanding contribution in reviewing, from the editors of *Applied Soft Computing Journal*, Elsevier.

Prof. Dr. Mangey Ram received the Ph.D. degree major in Mathematics and minor in Computer Science from G. B. Pant University of Agriculture and Technology, Pantnagar, India. He has been a Faculty Member for around 12 years and has taught several core courses in Pure and Applied Mathematics at undergraduate, postgraduate and doctorate levels. He is currently the *Research Professor* at Graphic Era (Deemed to be University), Dehradun, India. Before joining the Graphic Era, he was a Deputy Manager (Probationary Officer) with Syndicate Bank for a short period. He is Editor-in-Chief of *International Journal of Mathematical, Engineering and Management Sciences*, *Journal of Reliability and Statistical Studies,* Editor-in-Chief of six Book Series with Elsevier, CRC Press-A Taylor and Frances Group, Walter De Gruyter Publisher Germany, River Publisher and the Guest Editor &

Member of the editorial board of various journals. He has published 225 plus research publications (journal articles/books/book chapters/conference articles) in IEEE, Taylor & Francis, Springer, Elsevier, Emerald, World Scientific and many other national and international journals and conferences. Also, he has published more than 50 books (authored/edited) with international publishers like Elsevier, Springer Nature, CRC Press-A Taylor and Frances Group, Walter De Gruyter Publisher Germany, River Publisher. His fields of research are Reliability Theory and Applied Mathematics. Dr. Ram is a Senior Member of the IEEE, Senior Life Member of Operational Research Society of India, Society for Reliability Engineering, Quality and Operations Management in India, Indian Society of Industrial and Applied Mathematics. He has been a member of the organising committee of a number of international and national conferences, seminars and workshops. He has been conferred with "*Young Scientist Award*" by the Uttarakhand State Council for Science and Technology, Dehradun, in 2009. He has been awarded the "*Best Faculty Award*" in 2011; "Research Excellence Award" in 2015; and recently "*Outstanding Researcher Award*" in 2018 for his significant contribution in academics and research at Graphic Era Deemed to be University, Dehradun, India.